On Can

and the Transfinite

An Investigative Essay

C. J. Date

Published by:

TECHNICS PUBLICATIONS

TECHNOLOGY / LEADERSHIP

115 Linda Vista, Sedona, AZ 86336 USA
https://www.TechnicsPub.com

Cover design by Lorena Molinari.

First Printing 2023.

Printed in the United States of America.

ISBN, print ed.	9781634623278
ISBN, Kindle ed.	9781634623285
ISBN, ePub ed.	9781634623292
ISBN, PDF ed.	9781634623308

Library of Congress Control Number: 2023931249

When transcendental questions are under discussion,
be transcendentally clear.
—René Descartes (1596-1650)

I believe, and hope,
that future generations will laugh at this hocus pocus.
—Ludwig Wittgenstein (1889-1951),
referring to Cantor's diagonal argument

———— ♦♦♦♦♦ ————

To my tutors and mentors at Cambridge University,
especially Derek Taunt (1917-2004) of Jesus College,
who first taught me how it's possible to argue rigorously
about the infinite

About the Author

C. J. Date is an author, lecturer, researcher, and consultant, specializing in relational database technology. He was employed by LEO Computers in England from 1962 to 1967; by IBM in England and subsequently California from 1967 to 1983; and has been self-employed ever since, based first in California and latterly (from 2020 on) Vermont. He is best known for his book *An Introduction to Database Systems* (8th edition, Addison-Wesley, 2004), which has sold close to a million copies at the time of writing and is used by several hundred colleges and universities worldwide. He is also the author of numerous other books on database management. He was inducted into the Computing Industry Hall of Fame in 2004. He enjoys a reputation second to none for his ability to explain complex technical subjects in a clear and understandable fashion.

Contents

Abstract: In the late 1800s Georg Cantor (1845-1918) published a series of startlingly original papers that together laid the foundation for mathematical set theory. Here are some of the things he did in those papers:

1. He defined a set to be *countably infinite* (*countable* for short) if and only if it could be put into one to one correspondence with **N**, the set of all natural numbers.

2. He labeled that particular infinity—i.e., the cardinality of **N**—$\aleph_0$, where "$\aleph$"' (pronounced *aleph*) is the first letter of the Hebrew alphabet. $\aleph_0$ is the first *transfinite cardinal*, also known as the first *aleph number*.

3. He claimed that **R**, the set of all real numbers (reals for short), can't be put into one to one correspondence with **N** and is thus not countable; instead, it's *uncountably infinite* (*uncountable* for short). He labeled this new infinity—i.e., the cardinality of **R**—$\aleph_1$, and claimed that $\aleph_1$ is strictly greater than $\aleph_0$ (i.e., there are strictly more reals than there are natural numbers).

4. He then went on to claim that there exist infinitely many such transfinite cardinals ($\aleph_0$, $\aleph_1$, $\aleph_2$, ...), and moreover that those cardinals are totally ordered (i.e., $\aleph_0 < \aleph_1 < \aleph_2$ ...).

Today these and various related results, and the methods Cantor used to obtain them, are widely accepted; indeed, they're considered part of the foundation of modern logic and mathematics. Yet there are reasons to doubt them. The present essay, which is written for the nonspecialist, consists of a critical examination of such matters. Part I argues, first, that Cantor's famous diagonal proof of the uncountability of the reals is flawed; second, and more fundamentally, that his *method* of proof (i.e., diagonalization) is flawed as well. Part II then suggests that his earlier proof of the same result—which is based on intervals, and is less well known than his diagonal proof—is equally flawed. Finally, Part III describes a different way of looking at these matters, and concludes, not only that Cantor's proofs and proof methods are indeed flawed, but also that the results he obtained by them are in fact false.

Note: The seriousness of these matters can't be overstressed. To repeat, the proofs and proof methods in question are widely considered to be part of the foundation of modern logic and mathematics; so if they're flawed, there are huge implications. Every result that depends on them will need to be reexamined.

Acknowledgments: This essay has been through numerous iterations and has benefited hugely from the work of others, including in particular (but not limited to) that of Wolfgang Mückenheim, Erdinç Sayan, Robert White, and most especially David McGoveran. I'd also like to thank my critics and reviewers—some but not all of whom were persuaded by the arguments contained herein—for their many helpful comments.

Part I:

Cantor's Diagonal Argument

No one shall drive us out of the paradise that Cantor has created for us.
(Aus dem Paradies, das Cantor uns geschaffen,
soll uns niemand vertreiben können.)

—David Hilbert:
"Über das Unendliche" ("On the Infinite"),
Mathematische Annalen 95 (1926)

This tale grew in the telling. I first encountered Cantor's famous diagonal proof of the uncountability of the real numbers in 1958 when I was age 17 and still at grammar school, which was when I read the wonderful book *Mathematics and the Imagination*, by Edward Kasner and James Newman.[1] That book made a huge impression on me—especially Chapter 2 ("Beyond the Googol"), which is the chapter that describes Cantor's work. I remember finding that chapter pretty mind-boggling at the time, but I accepted it uncritically, and indeed continued to do so for many years thereafter.

What made me begin to question such matters was reading, in 2018 or so, a paper—still not widely published as of this writing—by my friend David McGoveran that did the same thing (question such matters, I mean).[2] I wrote a brief note at the time airing my thoughts on that paper and sent it to various friends (including McGoveran himself, of course) to get their reactions, and I've incorporated that note and some of the comments I received on it into what follows. But what really inspired me to tidy up those thoughts and turn them into an entire essay wasn't McGoveran's paper as such; rather, it was coming across another description of Cantor's work, this time in Marcus du Sautoy's book *The Great Unknown: Seven Journeys to the Frontiers of Science*,[3] which I read in 2020. That account, it seemed to me, left rather a lot to be desired, and I didn't want to let it go unchallenged.

By way of further background, I should explain that what I'm now calling just Part I was originally the essay in its entirety. But as I've said, I found that what I needed to say grew in the telling, and I eventually decided to split the essay overall into three separate parts. However, my main reason for doing so

---

[1] Originally published in 1940 by Simon and Schuster; republished by Dover Books in 2001.

[2] "Diagonalization," copyright © 2016-2021 Alternative Technologies, all rights reserved. I'll have a lot more to say about this paper at various points in what follows. *Note:* McGoveran has since split this paper into two parts and published them separately as preprints on *www.academia.edu*, under the titles "Cantor's Diagonal Argument: Three Refutations" and "Interval Arguments: Two Refutations of Cantor's 1874 and 1878 Arguments." But "Diagonalization" was what it was called when I first read it in 2018.

[3] Penguin Books, 2016.

was merely one of digestibility; the parts are still meant to be read as a whole (and in sequence), and they're definitely not meant to stand on their own.

I'd also like to make it clear that what follows isn't meant as any kind of research contribution. Rather, it's a narrative or popular account—I have a story to tell—and my target audience is educated lay readers, not mathematicians or logicians (at least, not primarily). I'm explicitly *not* making any claims of originality; all I'm doing is describing and commenting on what various other parties have had to say in connection with these matters.

One last preliminary point, a point of nomenclature. I referred earlier to "Cantor's famous diagonal proof." However, I'll show in what follows that the proof in question is flawed, and (worse) that the flaws are fundamental and can't be fixed. As a consequence, the "proof" isn't really a proof at all! So what can we call it? Cantor's "proof," in quotes? Cantor's argument? Cantor's claim? Cantor's would-be proof? Well, as you can see I opted for "argument" in my overall title for this part of the essay, and "Cantor's argument" actually isn't too bad; but most of the time I'll stay with the more familiar "Cantor's proof" (sometimes more specifically "Cantor's diagonal proof"), and leave it to you to interpret those words appropriately, depending on context.

CANTOR'S DIAGONAL PROOF

As noted above, du Sautoy's book *The Great Unknown: Seven Journeys to the Frontiers of Science* contains a discussion, on pages 400–401, of Cantor's famous diagonal proof to the effect that the real numbers ("the reals" for short) aren't countable—meaning, more precisely, that it's not possible to establish a one to one correspondence from the reals to the natural numbers (there are always some reals left over, as it were). Given this result, it follows that the cardinality of the reals must be strictly greater than that of the natural numbers. In other words, loosely: There are more reals than natural numbers.

> *Aside:* A few points of clarification are needed immediately. First, the *natural numbers*, also known as the counting numbers, are just the positive integers 1, 2, 3, etc. (*positive* meaning greater than zero, by definition). Some writers additionally consider zero to be a natural number; the literature isn't consistent on this point, but most writers, and indeed common usage, do exclude the zero case. To avoid this slight ambiguity, when I do want to include the zero case I'll use the term "nonnegative integers" instead of "natural numbers" (*nonnegative* meaning not less than zero—which is to say, greater than or equal to zero—again by definition).
>
> Second, I'm going to assume that the real numbers we're talking about are also all nonnegative; in other words, I'm going to ignore reals less than zero (barring explicit statements to the contrary, of course). The advantage of making this assumption is that it simplifies some of the arguments we're going to be examining, without at the same time materially affecting any of the conclusions to be drawn from those arguments.
>
> Third, I'm always a little leery of the term *one to one correspondence* (also of *one to many correspondence* and other such terms) because—as I've

explained at somewhat excruciating length elsewhere[4]—they're open to a variety of different interpretations However, in the present essay I'm just going to assume they're well understood. At least until further notice! *End of aside*.

Back to du Sautoy. Personally, I find du Sautoy's text in connection with these matters far too woolly to want to quote it verbatim. However, I believe the following paraphrase captures the essence of what he has to say without doing too much violence to his original text:[5]

> What Cantor does is build a real number with an infinite decimal expansion[6] that is guaranteed not to be on any given countable list of such numbers expressed in decimal notation. Suppose we have such a list. Then Cantor chooses the integer appearing in each decimal place of this new number one step at a time, working his way down the list. In the first decimal place, Cantor chooses an integer that's different from the integer in the first decimal place of the first number in the list. In the second decimal place, he chooses an integer that's different from the integer in the second decimal place of the second number in the list. Cantor carries on building this new number in this manner; that is, in the nth decimal place, he chooses an integer that's different from the integer in the nth decimal place of the nth number in the list. Consequently, the new number is different from every number in the list and thus doesn't itself appear in the list. That means there are numbers missing from the list.
>
> There are a few technical details to watch out for, but the essence of the foregoing argument is to show that there are more numbers with infinite decimal expansions than there are whole numbers. Cantor himself was genuinely surprised by his discovery: "I see it, but I do not believe it," he said.[7]

Regarding those numbers in the given list, incidentally, it turns out that, for reasons to be explained in the next section (and indeed as the foregoing extract tacitly assumes), we don't need to worry about digits to the left of the decimal point. In other words, we can limit our attention to just the fractional portions of the numbers in question. So we can think of that given list as an array in which

---

[4] In Chapter 8 of my book *Logic and Relational Theory: Thoughts and Essays on Database Matters* (Technics Publications, 2020), to be specific. PS: As the title of this reference might suggest, the various different interpretations I'm complaining about certainly do occur in the database world. They might or might not occur in the world of mathematics. But I'm not assuming my readers are mathematicians.

[5] One of my reviewers objected, not without justification, to my giving du Sautoy so much attention, especially since I find such a lot to criticize in what he has to say. But this is "an investigative essay"; it's meant for the lay reader, and, sadly, du Sautoy's presentation is all too typical of what the lay reader is likely to encounter elsewhere in this connection (further examples provided on request). In any case, it was specifically du Sautoy's presentation that got me motivated to write this essay in the first place, as I've said. PS: I note for the record that du Sautoy is the "Simonyi Professor for the Public Understanding of Science at the University of Oxford" in England. Make of that what you will.

[6] Du Sautoy's text uses decimal, as do most popular presentations of Cantor's work, but Cantor himself didn't—see the next section, also Appendix C to this part of the essay. *Note:* The term *expansion* as used by du Sautoy here could be criticized as well. However, it's fairly standard. E.g., Wolfram MathWorld: "The *decimal expansion* of a number is its representation in base 10 (i.e., in the decimal system)."

[7] Actually Cantor made this remark—*je le vois, mais je ne le crois pas*—in an 1877 letter to Richard Dedekind in connection with another of his findings: viz., that there are just as many points on the line forming one side of a given square as there are points in the interior of the square in question. (Personally, I regard this latter finding as comparatively reasonable: much more so than the suggestion that such things as "uncountable infinities" and "transfinite cardinals"—a concept to be explained later in this essay—might actually exist. For further elaboration of this position on my part, please read on.)

the ith row contains the ith number in the list, represented in decimal form, and the jth column corresponds to the jth decimal place within those numbers with respect to that decimal representation. Thus, the process of "building a new number"—i.e., building a number that, for all n, differs in the nth decimal place from the nth number in the list—can be thought of as:

a. Picking out the diagonal of that array that starts at top left, and then

b. Replacing each integer in that diagonal by some specific different integer, thereby producing what's sometimes called the corresponding *anti*diagonal.

Consequently, the process overall is usually referred to as *Cantor's diagonal method* (or *scheme*), or just *diagonalization* for short.

In this part of the essay I'll take a closer look at these matters. To be specific, I'll describe:

- First, what seems to me to be a valid refutation of Cantor's argument— namely, the one in the paper by McGoveran already mentioned;

- Second, another such refutation, also seemingly valid, by another friend, Robert White;

- Third, objections to those refutations from certain critics, not named here, with some responses to those objections by myself.

In addition I'll discuss an important theorem, also due to Cantor, to the effect that the cardinality of any given set is strictly less than that of the power set of the set in question.

McGOVERAN'S REFUTATION

Note: McGoveran's diagonalization paper actually contains three distinct refutations of Cantor's argument, only the first of which I describe in the present section. Also, the simplifications I discuss in the next two paragraphs are simplifications only as far as du Sautoy's presentation is concerned—both Cantor's original argument and McGoveran's refutation of that argument in fact already incorporate them.

As agreed, let's consider just the fractional part of the numbers we're talking about; in other words, let's consider just strings of digits, which we interpret as representing real numbers x in the range $0 \leq x \leq 1$.[8] Note that this simplification

_____

[8] You might be surprised to see that "$x \leq 1$" here, not "$x < 1$"—but if we allow those strings of decimal digits to be infinitely long (as indeed we do need to do when we start getting into details), then one such string will consist of all nines, and 0.9999..., where the dots mean "repeated forever," is a legitimate decimal representation of the value 1. (Though I note for the record that McGoveran disagrees with this latter assertion, and his refutation therefore takes x to be strictly less than one—i.e., he deals with the range $0 \leq x < 1$. I'll come back to this issue of numbers vs. their representations in Part III of this essay.)

is at least reasonable, in the sense that if we can show that the set of reals x in the range $0 \leq x \leq 1$ is uncountable, then it'll certainly follow that the set of *all* reals is uncountable.

Second, let's switch from decimal to binary. One advantage of this switch is that it eliminates the slight degree of arbitrariness involved in that business of "building new numbers," because now the process of choosing one of the integers 0, 1, ..., 9 for the nth decimal place—more precisely, the process of choosing a decimal integer for that nth decimal place that's different from the decimal integer in the nth decimal place of the nth decimal number in the list—is replaced by the following simpler procedure:

> For the nth binary place (i.e., the nth bit from the left), choose:
> 0 if the nth bit from the left of the nth number in the list is 1;
> 1 if the nth bit from the left of the nth number in the list is 0.

I'll refer to this procedure from this point forward as *the choice algorithm*.

> *Aside:* I put the phrase "building new numbers," and other phrases like it, in quotes in this essay because I subscribe to the view that, conceptually speaking, all numbers simply exist—there's no such thing as a "new" number, and thus there can't be any possibility of "building" such a thing.
> Let me elaborate on this point briefly. The root of the problem is that we usually don't bother to distinguish between (a) numbers as such, on the one hand, and (b) representations of numbers, using some specific base or radix, on the other. (Indeed, Cantor himself fails to make that distinction in his diagonal proof.) Thus, what's really meant by the phrase "building a new number" is merely a matter of somehow picking out, or identifying, the particular number in question—and we do that by spelling out a representation of that number in terms of some specific base or radix. (As already mentioned in footnote 8, this logical difference between numbers and their representations is something I'll be revisiting—indeed, I'll have quite a lot to say about it—in Part III of this essay.) *End of aside.*

Now to the substance of Cantor's argument. I'll begin by considering a small, and very obviously finite, analog of that argument. Suppose for the moment that we're limited to strings of length four bits. Here's a complete list of all such strings in numerical order:

```
0  0  0  0
0  0  0  1
0  0  1  0
0  0  1  1

0  1  0  0
0  1  0  1
0  1  1  0
0  1  1  1

1  0  0  0
1  0  0  1
1  0  1  0
1  0  1  1
```

```
1  1  0  0
1  1  0  1
1  1  1  0
1  1  1  1
```

However, I don't want the argument that follows to rely, or even appear to rely, on this obvious numeric ordering—nor on any other particular ordering, come to that—so let me shuffle the strings into some arbitrarily different sequence. Here's one possible reshuffling (and I'll focus on this one in what follows, just for definiteness):

```
 1.  1  0  0  0
 2.  1  1  1  1
 3.  0  1  0  1
 4.  1  0  0  1

 5.  1  1  1  0
 6.  0  0  1  1
 7.  0  1  1  1
 8.  1  1  0  1

 9.  0  0  0  0
10.  1  0  1  0
11   0  1  0  0
12.  0  1  1  0

13.  0  0  1  0
14.  1  0  1  1
15.  0  0  0  1
16.  1  1  0  0
```

As you can see, I've labeled the entries in this reordered list with (decimal) entry numbers 1-16 for purposes of subsequent reference. Note also that labeling the entries in this way is of course logically equivalent to counting them.

Now I want to focus on just the first four entries in this list—four, because we're considering strings of length four bits and we need to talk about the diagonal (more specifically, we want to be able to *diagonalize*):

```
 1.  1  0  0  0
 2.  1  1  1  1
 3.  0  1  0  1
 4.  1  0  0  1
```

I've highlighted the diagonal because, as I've said, I now want to diagonalize; that is, I want to use the choice algorithm to "build a new number" that differs from the first number in the list in the first position, from the second number in the list in the second position, and so on. Here then is that "new number" (note that it's uniquely defined, thanks to our decision to simplify by switching from decimal to binary):

```
0  0  1  0
```

This number is clearly different from every number in the first four entries in the list—necessarily so, because of the way we've "built" it. Equally clearly, though, it's *not* different from every number in the remainder of the list; to be specific, it's identical to the number in position 13.

Observe now that the complete list of 4-bit numbers is deeper than it's wide—it's four bits wide, of course, but it's $2^4 = 16$ entries deep. In fact, it should be obvious that, for any given $n \geq 1$, (a) the complete list of n-bit numbers is n bits wide but 2^n entries deep, and further that (b) 2^n is greater than n—in fact, 2^n grows much faster than n does (as is well known, of course), because 2^n grows exponentially while n grows only linearly.

Aside: It's easy to see that $2^n > n$ for all $n \geq 1$. Here's an inductive proof:

    <u>Basis:</u> Let $n = 1$. Then $2^n = 2^1 = 2$, and $2 > 1$.

    <u>Inductive step</u> (assume that $2^n > n$ for some $n \geq 1$, and show that it follows that $2^{n+1} > n+1$):

        $2^{n+1} = 2 \times 2^n = 2^n + 2^n$.
        So $2^{n+1} > n+n$ (because by our assumption $2^n > n$).
        Hence certainly $2^{n+1} > n+1$ (because by our assumption $n \geq 1$).

    <u>Conclusion:</u> $2^n > n$ for all $n \geq 1$.[9]

End of aside.

It follows that for any $n \geq 1$ the complete list of n-bit numbers can be divided into two disjoint sublists:

a. The *initial square* sublist, consisting of the first n entries (n bits wide and n entries deep), followed by

b. The *remainder* sublist, consisting of the remaining 2^n-n entries (n bits wide and 2^n-n entries deep).

Observe now that diagonalization applies *by definition* only to the initial square sublist. That is, for any given n the choice algorithm yields a number that *by definition* doesn't appear in the first n entries in the list (i.e., in the initial square sublist), but does necessarily appear in the remaining 2^n-n entries (i.e., in the remainder sublist). Note too that, since 2^n is always strictly greater than n, the remainder sublist is never empty; in fact, it very quickly dwarfs the initial square sublist, meaning that "almost all" of the numbers that appear in the list at all actually appear in the remainder sublist. As a consequence, the fact that diagonalization fails to produce an entry in the initial square sublist becomes less and less surprising, as it were, as n increases.

What's not clear to me is why, as Cantorians—if I might be allowed to use such an expression—seem to believe, the foregoing argument should cease to be

---

[9] Actually, since $2^0 = 1$ and $1 > 0$, we have the stronger result that $2^n > n$ for all $n \geq 0$ (instead of just $n > 0$ as above). But the case $n = 0$ isn't very relevant as far as this essay is concerned.

valid as n goes to infinity. No matter how large n might be, the list still divides into an $n \times n$ initial square sublist and an $n \times (2^n - n)$ remainder sublist, and the diagonal argument clearly applies (as the very term "diagonalization" itself strongly suggests) only to the initial square sublist (once again: *by definition*).

> *Aside:* I'd like to add a couple of postscripts to the foregoing argument, both of them due like that argument itself to McGoveran:
>
> ■ First, as we've seen, the remainder sublist contains $2^n - n$ more entries than the initial square sublist. But if we'd worked in decimal instead of binary, it would have contained $10^n - n$ more entries! This state of affairs shows rather clearly that Cantor's argument has to do with representations of numbers and not with numbers as such.
>
> ■ Second, it's important to understand that the foregoing argument isn't just a refutation of diagonalization as a basis for "proving" that the reals are uncountable in particular—*it's a refutation of diagonalization as a method of proof in general.* In other words, *any* proof that relies on diagonalization in some shape or form becomes suspect at best.[10]
>
> *End of aside.*

WHITE'S REFUTATION

Note: McGoveran's diagonalization paper also includes a discussion, predating White's work, of something equivalent to what I'm referring to here as White's refutation. However, White came up with his argument independently.

Here again are the first four entries in our list of strings of length four bits (i.e., take n to be four again, just for the moment):

1.	1	0	0	0
2.	1	1	1	1
3.	0	1	0	1
4.	1	0	0	1

Now, however, let's interpret the strings differently: Instead of thinking of them as we did before as denoting real numbers x in the range $0 \le x \le 1$, let's think of them as denoting nonnegative integers x—specifically, integers x in the range $0 \le x \le 15$, since we're limiting our attention for the moment to 4-bit strings. Thus, the four strings shown represent, in top to bottom order, the decimal integers 8, 15, 5, and 9, respectively.

---

[10] Interestingly, a case in point—a case, that is, of a proof that relies (perhaps not very obviously) on diagonalization—is Gödel's proof of his First Incompleteness Theorem. What that theorem says is this: If F is a consistent formal system that's at least as powerful as elementary arithmetic, then F is incomplete, in the sense that there exist statements in F that are true but can't be proved in F. So it seems that this famous result might also need to be examined further ... Could Gödel be wrong?

Next, let's apply the choice algorithm to "build a new number"—more specifically, to "build a new *integer*." This time, however, instead of using the diagonal from top left to bottom right as we did before, we need to use the one from top right to bottom left. The algorithm becomes:

For the *n*th bit from the right, choose:
0 if the *n*th bit from the right of the *n*th number in the list is 1;
1 if the *n*th bit from the right of the *n*th number in the list is 0.

Here again are the first four entries in the list, with the positions on the diagonal from top right to bottom left highlighted:

1. 1   0   0   *0*
2. 1   1   *1*   1
3. 0   *1*   0   1
4. *1*   0   0   1

So the diagonal is

0   1   1   1

As for the "new integer"—well, this part of the argument can be a little confusing, so let me take it very carefully, one step at a time:

■ The first bit from the right in the first entry in the list (which becomes the first bit from the *left* in the diagonal) is 0, so the first bit from the right in the "new integer"—the antidiagonal—is 1.

■ The second bit from the right in the second entry in the list (which becomes the second bit from the left in the diagonal) is 1, so the second bit from the right in the "new integer" is 0.

And so on. Thus, the "new integer" is

0   0   0   1

(decimal one). This integer clearly doesn't appear in the first four entries in the list. Equally clearly, however, it does appear in the rest of the list; to be specific, it's identical to the integer in entry number 15. But an argument precisely analogous to Cantor's original argument would say that when *n* goes to infinity, then that "new integer" never appears in the list at all, and hence that the *nonnegative integers* 0, 1, 2, 3, ... aren't countable!—which is, I hope it goes without saying, absurd, and in fact a contradiction in terms.

Aside: One critic of an early draft of this essay objected at this point that the choice algorithm breaks down when the entries in the list become infinitely long, because "the *n*th bit from the right" isn't defined for a string of infinite length. But all talk of "left" and "right" in contexts like the one at hand is merely a matter of convention: a convention, that is, as to how we choose to write the strings in question, and how we then go on to interpret them.

I'd like to elaborate on this point briefly—but before I do, let me say that (a) you might be excused for finding the explanation that follows a little confusing, and (b) you won't miss too much if you decide to overlook it. Nothing in the remainder of this essay depends on it.

That said, let's agree to stay with four bits for simplicity. Let the bit string $t_4t_3t_2t_1$ represent some nonnegative integer in conventional binary notation. Then moving right to left along this string corresponds to increasing powers of two; for example, the string 1011 denotes $(1\times2^0)+(1\times2^1)+(0\times2^2)+(1\times2^3)$, or in other words decimal eleven. Suppose we now reverse that string—i.e., write it "the other way around," as it were, as $t_1t_2t_3t_4$. All that happens is that now it's moving left to right along the string that corresponds to increasing powers of two. For example, reversing the string 1011 yields 1101, which—now reading left to right—again denotes $(1\times2^0)+(1\times2^1)+(0\times2^2)+(1\times2^3)$, which of course is still decimal eleven.

Also (and still assuming that our 4-bit strings are to be understood as representing nonnegative integers): With conventional notation the string $t_4t_3t_2t_1$ must be understood as having an infinite number of leading zeros; with the reverse notation described above, the string $t_1t_2t_3t_4$ must be understood as having an infinite number of trailing zeros instead. However, there's no need to consider any of these zeros when we diagonalize (see below), since we're explicitly restricting our attention to strings of length 4 bits. Indeed, essentially the same is true for strings of any finite length, since all we're doing in every such case is, in effect, limiting our attention to the initial square sublist.

So what happens in our example if we reverse notation as described above? The first four strings become

```
1.  0   0   0   1
2.  1   1   1   1
3.  1   0   1   0
4.  1   0   0   1
```

(but are still interpreted as decimal 8, 15, 5, and 9, respectively, as before). The choice algorithm now reverts to its original form—i.e., it starts at top left instead of top right. The diagonal in the example is thus still 0111, and the "new integer" (the antidiagonal) is accordingly 1000; but that "new integer" must be understood as representing decimal 1 (as before), and it appears as entry 15 in the list (as before)—because, to say it again, the entries in that list have now all been reversed. *End of aside.*

I'd like to take a moment to stress the absurdity of the foregoing conclusion (viz., that the set of nonnegative integers is uncountable). After all, to say the nonnegative integers are uncountable is to say they can't be put into one to one correspondence *with themselves*.

Let's pause for a moment to take stock of where we are. Cantor claimed to have proved that the reals are uncountable; what's more, his proof seems to be very widely accepted. However, I believe the arguments of McGoveran and White demonstrate rather clearly that the proof in question is flawed, and in fact no proof at all. Note carefully, however, that we can't reasonably conclude from this state of affairs that the reals are countable—all we can conclude is that the diagonal argument doesn't prove they aren't.

RESPONDING TO CRITICISM

I showed an early draft of this essay up to this point to several mathematically inclined friends and friends of friends, and received detailed criticisms from two of the parties concerned. In this section, I'd like to respond to some of what those critics had to say. *Note:* The critics in question were both male, which accounts for my use of male pronouns in what follows.

First Critic

Here's the essence of what the first critic said (I'm quoting his text essentially verbatim, except that I've omitted a few minor portions that aren't relevant to the matter at hand):

> Your venture into a finite analog of Cantor's argument creates, I believe, a number of straw men that don't appear in Cantor's proof. They take us away from examining the essence of his argument, which is splendidly simple:
>
> 1. Let's assume that there's at least one algorithmic way of generating an infinite list S of all the reals such that the list is in one to one correspondence with the list P of positive integers.[11] [*This critic explicitly framed his objections in terms of positive integers, not natural numbers and not nonnegative integers.*]
>
> 2. Start writing down the corresponding elements of S [*which the critic implicitly assumes contains no duplicates*] against the elements of P starting with 1 and increasing by 1 at each step.
>
> 3. No matter how far we go down that list, I can always write down a real X that doesn't appear in the list so far.[12] I can keep [appending] the next binary or decimal place of X when I'm [giving] the next element of S in such a way that this statement remains true.
>
> 4. Therefore there'll always be at least one real X that can't be paired with a positive integer. [*Sorry, but this simply doesn't follow. To put the point more positively, we can simply pair that real X with the next available integer. See Appendix B to this part of the essay for a striking illustration of this point.*]

---

[11] McGoveran commented on this first point as follows: "Already we see the implicit and unacknowledged assumption that one to one correspondence with such a list [*i.e., between list S and list P*] is necessary (rather than merely sufficient) to show countability." See Part III of this essay for further discussion. *Note:* List S here is an example of what later I'll be calling a *Cantor list*.

[12] McGoveran commented on this claim also: "And this is the key point: only *so far*. The fact that there are more and more numbers in the remainder sublist that are forever unreachable by the diagonal argument makes this point a fatal one."

5. But that contradicts the assumption in Step 1, which is therefore false, and the elements of S can't be put into one to correspondence with the elements of P. So the set of reals is by definition uncountable.

End of Cantor's proof. No mention of finite subsets or initial square / remainder sublists.

Wherein lies the fallacy in [the foregoing]?

My response to this critic consisted of a few generalities, followed by a more detailed analysis of the critic's five-step argument. Here first are the generalities (though I've tidied them up somewhat here):

■ First, I want to make it clear that I'm not claiming—at least, not in what I've said in this essay so far—that Cantor's result (that the reals are uncountable) isn't valid. What I *am* claiming is that the diagonalization scheme per se isn't valid, meaning it doesn't do what Cantor and so many other people seem to think it does. And if I'm right on this, then all results produced by that scheme become suspect.

■ The critic takes me to task for introducing "straw men" (straw persons?). But I don't believe they *are* straw men; I believe they can provide some helpful intuitive understanding of the thinking behind diagonalization, and they can help the reader see why the diagonal proof doesn't work.

■ The critic then offers what he calls "the essence of [Cantor's] argument." But I don't agree that what he shows is that essence. The "essence" of the argument is diagonalization, and he doesn't even mention that—not as such, at any rate. Though I suppose Step 3 of the critic's five-step procedure is an attempt to explain Cantor's process of "building a new number," and must therefore be understood as being equivalent, somehow, to Cantor's diagonalization scheme.

I then went on to consider that five-step procedure more carefully. In fact, what I did was present a slightly modified version of that procedure (shown below in italics), using the critic's own wording as much as possible. My changes to that wording are indicated by showing the critic's original text with double strikethrough and then immediately following that original text with replacement text in bold.

*1. Let's assume that there's at least one algorithmic way of generating an infinite list S of all the ~~reals~~ **positive integers** such that the list is in one to one correspondence with the list P of positive integers.*

Comment: Of course, replacing *reals* in this first step by *positive integers* as I've just done does make it rather easy to find "an algorithmic way of generating" that infinite list *S*!

2. Start writing down the corresponding elements of S against the elements of P starting with 1 and increasing by 1 at each step.

Comment: OK—S and P are now both just lists of all the positive integers, so a one to one correspondence can obviously be established between them, as indeed the critic assumes in his Step 1. Of course, the simplest such correspondence is rather trivial—just pair each integer with itself.

3. No matter how far we go down that list, I can always write down a ~~real~~ **positive integer** *X that doesn't appear in the list so far. I can keep appending the next binary or decimal place of X when I'm giving the next element of S in such a way that this statement remains true.*

Comment: Assuming conventional positional notation, "appending the next binary or decimal place" now has to be understood, in my positive integers revision, as appending at the left, not the right.

4. Therefore there'll always be at least one ~~real~~ **positive integer** *X that can't be paired with a positive integer.*

Comment: This revised version of Step 4 is self-evidently false.

5. But that contradicts the assumption in Step 1, which is therefore false, and the elements of S cannot be put into one to correspondence with the elements of P. So the set of ~~reals~~ **positive integers** *is by definition uncountable.*

Comment: The revised conclusion here is self-evidently false as well (indeed, in the critic's own words, it's false "by definition"). In fact, the only thing the foregoing argument shows—regardless of whether we're talking about the critic's original argument or my revised version of that argument—is that, no matter how far we go down the list, there'll always be at least one real number (in the critic's version) or positive integer (in my version) that we haven't reached yet. It does *not* follow that we'll never reach it, and what I'm arguing is that in fact we always will. That's what that business of the initial square sublist vs. the remainder sublist was all about.

    To put the point another way: If the critic's argument shows the reals are uncountable, then the foregoing analog of that argument shows the positive integers are uncountable as well. Since the consequent here is false, the antecedent must therefore be false as well.

Second Critic

Here now is what the second critic had to say. (Again I'm quoting essentially verbatim; I've made a few minor revisions for reasons of flow and the like, and I've added some comments in brackets, but I haven't changed the sense. Apart from those comments, italics are as in the original.)

Date is evidently missing an understanding and appreciation of the rigorous framework that mathematics has developed for dealing with the infinite, including the theory of infinite sequences and series and convergence thereof, which is generally regarded as one of its crowning achievements. [*Note: The difference in mathematics between a sequence and a series is as follows: A sequence is an ordered list of elements; a series is the sum of such a sequence—assuming, of course, that the elements of the sequence in question are such that the operation of addition makes sense for them. The distinction isn't important at this juncture.*]

As for his comments on Cantor's proof that the reals are uncountable, he seems to be missing its fundamental logic. As a proof by contradiction, it begins by assuming the opposite, viz., that the reals are countable. Now *by definition* this means that there exists an enumeration of the reals, i.e., a *one to one correspondence with the natural numbers*, i.e., *an infinite sequence*! So that's exactly what Cantor supposes and lays out—an infinite sequence that supposedly includes all real numbers, each of which he represents by an infinite digital expansion (as per the universally accepted characterization of real numbers). At this point he hasn't done anything more than assume the opposite of what he wants to prove.

Incongruously, Date refers to this [*i.e., laying out that infinite sequence*] as something that Cantor has "magically managed" to do "despite the impossibility of doing so."[13] By "impossibility" I presume he's referring to the impossibility of writing down all the digits of a given (nonterminating) real number, from which he's managed to infer ("magically," I'd have to say) that the core construct of Cantor's proof is invalid even before he launches into his diagonalization process.

Date seems to be beset with a basic confusion whereby he leaps from the fact that we can't write down an infinite number of digits to the conclusion that we can't thereby work with the sequence in question. But this is the "magic" of mathematics, if you like—that it's sufficient to have a formula or an algorithm for generating any given element, and thereby arbitrarily many elements, of a sequence, which in fact is how we *define* a countably infinite sequence. Given that there's no element of the sequence that we're unable to write down it follows that in fact we *can* write them all down!

Well, naturally I have some reactions to the foregoing criticisms. I'll take them one piece at a time.

Date is evidently missing an understanding and appreciation of the rigorous framework that mathematics has developed for dealing with the infinite, including the theory of infinite sequences and series and convergence thereof, which is generally regarded as one of its crowning achievements.

The critic is entitled to his opinion, of course, but I really don't think I'm guilty of the charge of "missing an understanding of mathematics" that I'm being accused of here. I'm a mathematician myself (at least, I used to be): never a very good one, as I'd be the first to admit, but a mathematician nonetheless. My degree is in mathematics. In particular I do understand how the mathematical theory of limits provides a rigorous means for dealing with infinite sequences and the like. Thus, I reject the suggestion that I find McGoveran's argument

---

[13] Yes, I'm afraid I did say something like this in my early draft. Further discussion to follow!

convincing only because I don't know enough about these matters to be able to spot the flaw (assuming there is one, of course, which—let me be clear—I don't think there is).

As for [Date's] comments on Cantor's proof that the reals are uncountable, he seems to be missing its fundamental logic.

I deny this charge. The "fundamental logic" of Cantor's proof is as follows:

1. Assume the reals are countable.

2. Then they can be put into one to one correspondence with the nonnegative integers.

3. So in principle there exists an infinite list of ordered pairs of the form $<i,r>$, where i is a nonnegative integer and r is a real number, such that there's exactly one pair in the list for each possible i and exactly one pair in the list for each possible r.

> *Aside:* As du Sautoy says in his own description of Cantor's diagonal argument, "there are a few technical points to watch out for," and here's one of them: namely, that (as already noted in footnote 8) certain reals have two distinct representations in terms of whatever base or radix we happen to be using. For example, the decimal strings 0.10000... and 0.09999... both denote the real "one tenth." So do we need to worry about the possibility that our list of pairs might contain two distinct pairs for the same real? The short answer is "No, we don't—at least, not much." A slightly longer answer might be "No, we don't, so long as those pairs contain numbers as such and not representations of numbers"—except that, in order for Step 4 below to make any sense, they must in fact contain representations after all. But even so, I still claim we don't need to worry about the issue very much. For the time being, therefore, let's agree to just forget about it; as I said a few pages back, I'll be discussing it in detail in Part III of this essay. *End of aside.*

4. Cantor then "builds" a certain real number that he says isn't in any pair in the list. The existence of such a real contradicts the assumption in Step 1, which must therefore be wrong. So the reals aren't countable.

My problem with this "fundamental logic" has to do with Step 4. It's my claim that all that Cantor shows is that the number he "builds" isn't in the initial square sublist—*not* that it isn't in the list anywhere at all.

As a proof by contradiction, it begins by assuming the opposite, viz., that the reals are countable. Now by definition this means that there exists an enumeration of the reals, i.e., a one to one correspondence with the natural numbers, i.e., an infinite sequence! So that's exactly what Cantor supposes and lays out—an infinite sequence that supposedly includes all real numbers, each of which he represents by an infinite digital expansion (as per the universally accepted characterization of real numbers). At this point he hasn't done anything more than assume the opposite of what he wants to prove.

Actually (and crucially) he *has* done something more—he's assumed that his "supposed, laid out, infinite sequence" takes the form of a square array and thus has a diagonal, an assumption that's easily seen to be unwarranted and false. Still, let's soldier on ... Clearly, the critic's text here corresponds to Steps 1-3 of the "fundamental logic" of Cantor's argument as I described it above. Though I might quibble with the critic over his use of the phrase "lays out"; "laying out" the list sounds rather as if he's making that abstract list concrete, which we all agree is something that can't actually be done. See the critic's next comment and my response to it.

Incongruously, Date refers to this as something that Cantor has "magically managed" to do "despite the impossibility of doing so." By "impossibility" I presume he's referring to the impossibility of writing down all the digits of a given (nonterminating) real number, from which he's managed to infer ("magically," I'd have to say) that the core construct of Cantor's proof is invalid even before he launches into his diagonalization process.

This criticism is valid. In my original comments on McGoveran's piece, I wrote the following (lightly edited here once again):

> Cantor shows that an attempt to list all the reals is futile. But this is obviously so, since there are real numbers—in fact an infinite number of them—that we can't write down in decimal form individually (never mind trying to write down all of them). So I don't see how Cantor's diagonal argument can warrant his conclusion, that there are real numbers not included in the list he has somehow magically managed to write down despite the impossibility of doing so. In other words, I don't see how that diagonal argument per se can be used to justify the conclusion that the reals are uncountable.

I see now that what I wrote in these comments was muddled, and I hereby apologize for the confusion. Though let me note for the record that I was correct in claiming that there are real numbers that can't be physically written down, and I was therefore correct a fortiori in claiming that a complete list of the reals can't be physically written down either. But Cantor's diagonal argument doesn't rely on the notion that such a list has to be physically written down—all it relies on is the idea that such a list might exist in the abstract.

That said, however, I still don't see how Cantor's argument can possibly be valid, for reasons explained earlier.

Date seems to be beset with a basic confusion whereby he leaps from the fact that we can't write down an infinite number of digits to the conclusion that we can't thereby work with the sequence in question. But this is the "magic" of mathematics, if you like—that it's sufficient to have a formula or an algorithm for generating any given element, and thereby arbitrarily many elements, of a sequence, which in fact is how we define a countably infinite sequence. Given that there's no element of the sequence that we're unable to write down it follows that in fact we can write them all down!

OK, I accept that I was "beset with a basic confusion" here. I still don't agree that Cantor's argument is valid, though—and I'm not sure about the critic's final sentence here, either. Being able to write down "a formula or an algorithm for generating any given element of a countably infinite sequence" might indeed mean we can write down any individual element; however, I'm not at all sure that being able to write down any individual element is the same as being able to write them all down.[14] After all, the first of these operations is finite while the second is infinite, and there thus seems to me to be a rather important logical difference between them.

CANTOR'S POWER SET THEOREM

I turn now to a different topic, albeit one that's directly related and relevant—in fact highly relevant—to the overall theme of this essay nonetheless. Let set s have finite cardinality n, and let S be the power set of s—i.e., the set whose elements are all of the subsets of s. For example, let s be the set $\{a,b,c\}$, of cardinality 3; then S is the set containing just the following (namely, all of the subsets of s and nothing else):

$$\{\} \quad \{a\} \quad \{b\} \quad \{c\} \quad \{a,b\} \quad \{b,c\} \quad \{c,a\} \quad \{a,b,c\}$$

As you can see, S in this example is of cardinality $2^3 = 8$. More generally, in fact, if s has cardinality n, then S has cardinality 2^n. (In particular, if s is the empty set $\{\}$, then S is the set $\{\{\}\}$, which has cardinality $2^0 = 1$.) And since (as we already know) 2^n is greater than n for all $n \geq 0$, it follows that S is always "bigger than" s, in the sense that it's always of strictly greater cardinality.

Actually, the foregoing result—viz., that S is always "bigger than" s, *even in the infinite case* (i.e., even if s is infinite), was proved by Cantor himself, and is now generally known as Cantor's Theorem. Using the standard notation $|s|$ to denote the cardinality of set s, one possible formulation of Cantor's proof is as follows (note that it falls into two parts):[15]

---

[14] Well, it demonstrably isn't. By the way, it's relevant here to mention the notion, due to Émile Borel and popularized by Gregory Chaitin, that certain numbers—to be specific, random numbers, which is to say numbers whose digits follow no predictable pattern, no matter what radix is used—are incapable of representation by any "formula or algorithm" that's shorter than the number in question, anyway. Indeed, that's Chaitin et al.'s *definition* of what it means for a number to be random in the first place. And given that definition, such numbers really can't be written down physically at all (not even by means of "a formula or an algorithm").

[15] As a matter of fact the proof is essentially, though perhaps not obviously, just another application of diagonalization, and for that reason is necessarily somewhat suspect. For further discussion of this point, see Part III of this essay.

I.   Show $|s| \leq |S|$.[16]

Consider the following *one to one into mapping M1* from s to S:[17]

For each element $x \in s$,[18] define the image of x under $M1$ to be the element $\{x\} \in S$.

This mapping is clearly well defined, and it implies among other things that $|s| \leq |S|$.

II. Show $|s| \neq |S|$.

Suppose $|s| = |S|$. Then there must exist a *one to one onto mapping M2* between s and S,[19] such that every element $x \in s$ maps under $M2$ to some unique element $X \in S$ and every element $X \in S$ maps under $M2$ to some unique element $x \in s$.

Since X is a set of elements of s, it must be the case that the unique corresponding element $x \in s$ either appears in X ($x \in X$) or it doesn't ($x \notin X$).

For all $X \in S$, let V be the set of elements $x \in s$ such that $x \notin X$.

By definition $V \subseteq s$,[20] and therefore $V \in S$. By our assumption, therefore, V is the image under $M2$ of some $v \in s$. Does $v \in V$?

(*Yes*) If $v \in V$, then v violates the set defining property for V, and so $v \notin V$.

(*No*) If $v \notin V$, then v satisfies the set defining property for V, and so $v \in V$.

Either way we have a contradiction. It follows that no such $M2$ exists, and hence that $|s| \neq |S|$.

---

[16] Please understand that "showing that $|s| \leq |S|$" is only a very informal way of characterizing what Part I of the proof actually does. After all, to suggest that the cardinality of one set might be less than that of another, if the sets in question are infinite, is to assume it makes sense to apply "<" to infinities. But such comparisons are legitimate only if the very result we're trying to prove is in fact the case! (A similar remark applies to Part II of the proof as well, of course.)

[17] A *one to one into mapping* (also known as an *injection*) from set $s1$ to set $s2$ is a mapping, or function, such that each element of $s2$ is the image under that mapping of at most one element of $s1$.

[18] The expression $x \in s$ means x is contained in s (equivalently, x is an element of s).

[19] A *one to one onto mapping* (also known as either a *strict one to one correspondence* or a *bijection*) from set $s1$ to set $s2$ is a mapping, or function, such that each element of $s2$ is the image under that mapping of exactly one element of $s1$. Note that if M is such a mapping, then it has an inverse mapping M' from $s2$ to $s1$ that's also one to one and onto, and the term *one to one onto mapping between* is often used—and is used in this essay—to refer to such a mapping M and its inverse M' considered in combination.

[20] The expression $V \subseteq s$ means V is included in s (equivalently, V is a subset of s).

Taken together, the results of Parts I and II of this proof show that $|s| < |S|$, and S is thus "bigger than" s. QED.

> *Aside:* Note the similarity—the family resemblance, one might say—between the set V as defined in Part II of the foregoing proof and the (impossible) "set of all sets that don't contain themselves as an element." This latter is, of course, the basis of Russell's famous paradox.[21] As a matter of fact, it has been suggested that Russell first came up with his paradox after studying Cantor's Theorem, and Cantor's proof of that theorem. *End of aside.*

UNANSWERED QUESTIONS

Regardless of whether you think Cantor's diagonal proof is successful in its attempt to show that the reals are uncountable—of course, I tried to show earlier that it isn't—it does seem to follow as a corollary from his theorem (i.e., "Cantor's Theorem" as discussed in the previous section) that there are an infinite number of "transfinite cardinals."[22] Why? Because if the (infinite) cardinality $|s|$ of an infinite set s is strictly less than the (also infinite) cardinality $|S|$ of its power set S, then $|S|$ in turn must be strictly less than the cardinality $|S'|$ of the power set S' of S, and so on ad infinitum (I choose my words carefully).

So does that corollary make sense? Well, here's the generally accepted definition (due to Dedekind) of what it means for a set s to be infinite:

> **Definition (infinite set):** Set s is infinite if and only if it can be put into a strict one to one correspondence with some proper subset p of itself—in other words, if and only if a "one to one onto" mapping can be defined between the elements of s and the elements of p.

For example, the set of nonnegative integers $\{0,1,2,3,...\}$ is infinite, because there's a strict one to one correspondence (or one to one onto mapping) between that set and the set of perfect squares $\{0,1,4,9,...\}$, and this latter set is clearly a proper subset of the former one. As a simpler example, there's also a strict one to one correspondence or one to one onto mapping between the nonnegative integers $\{0,1,2,3,...\}$ and the even nonnegative integers $\{0,2,4,6,...\}$, and again this latter set is clearly a proper subset of the former one.

> *Aside:* If the infinite set s in question happens to be a power set, what proper subset p of itself do you think it might be put into strict one to one correspondence with? You might like to ponder this question yourself for a while before reading any further.

---

[21] Russell's Paradox can be stated thus: Let W be the set of all sets that don't contain them themselves as an element. Does W contain itself? *Yes* and *no* both lead to a contradiction. *Note:* This paradox is always attributed to Russell, but (according to Wikipedia) other logicians and mathematicians—Zermelo, Hilbert, and others—were certainly aware of it before Russell articulated it in 1901.

[22] Cardinal numbers ("cardinals") are the counting numbers 1, 2, 3, etc. They're distinguished from the ordinal numbers ("ordinals") 1st, 2nd, 3rd, etc.

Here now is one possible answer (with acknowledgments to Jon Lenchner of IBM).[23] For definiteness, let s be, specifically, the power set of the set of nonnegative integers, and let p be what remains of s when all singleton sets— i.e., sets of the form $\{i\}$—are removed:

$$p \stackrel{\text{def}}{=} \{ x : x \in s \text{ AND } |x| \neq 1 \}$$

Clearly, p as just defined is a proper subset of s. Now consider the following mapping M from elements of s to elements of p:

- *Singletons:* For all i, map the element $\{i\}$ of s to the element $\{0,i+1\}$ of p:

```
{0} -> {0,1}
{1} -> {0,2}
{2} -> {0,3}
     etc.
```

- *Doubletons:* For all i and j, map the element $\{i,j\}$ of s to the element $\{i+1,j+1\}$ of p:

```
{0,1} -> {1,2}
{0,2} -> {1,3}
{0,3} -> {1,4}
     etc.

{1,2} -> {2,3}
{1,3} -> {2,4}
{1,4} -> {2,5}
     etc.
```

- *Others:* Map every other element x of s to itself:

```
x -> x
```

I'll leave to you to verify that the mapping M from s to p as just defined is indeed one to one and onto. (Note, therefore, that the inverse mapping M' from p to s, obtained from M as defined above by simply reversing the direction of the arrows, is one to one and onto as well.) *End of aside.*

To repeat, set s is infinite if and only if it can be put into a strict one to one correspondence with some proper subset p of itself. Observe, therefore, that this definition implies that $|p| = |s|$. Of course, given that p is a *proper* subset of s, we might have expected that $|p| < |s|$ (i.e., that $|p|$ is strictly less than $|s|$). The apparent contradiction involved here serves, I think, to raise some obvious questions regarding the propriety of applying comparison operators such as "<" to infinite comparands—which is what the proof, or at any rate the conclusion of that proof, of Cantor's Theorem most certainly does.

Aside: Actually the comparison operators "=", "<", etc., on transfinite cardinals are formally defined in terms of *mappings*, as follows:

---

[23] You might notice some similarities between Lenchner's solution and the well known parable (some call it a paradox) of Hilbert's Hotel. See *http://en.wikipedia.org/Hilbert's_paradox_of_the_Grand_Hotel.*

- The transfinite cardinal *n1* is defined to be *equal to* the transfinite cardinal *n2* ("*n1* = *n2*") if and only if (a) there exist sets *s1* and *s2* whose cardinalities |*s1*| and |*s2*| are *n1* and *n2*, respectively, and (b) there exists a one to one onto mapping between *s1* and *s2*—in which case those sets are said to be *equinumerous* (or *equipotent*, or *equipollent*).

- The transfinite cardinal *n1* is defined to be *less than or equal to* the transfinite cardinal *n2* ("*n1* ≤ *n2*") if and only (a) there exist sets *s1* and *s2* whose cardinalities |*s1*| and |*s2*| are *n1* and *n2*, respectively, and (b) *s1* is a subset of *s2*.

- The transfinite cardinal *n1* is defined to be *less than* the transfinite cardinal *n2* ("*n1* < *n2*") if and only *n1* ≤ *n2* is true and *n1* = *n2* is false.

Given the foregoing definitions of "=" and "<", incidentally, transfinite arithmetic turns out to be rather simple! To be specific, let at least one of the cardinal numbers *n1* and *n2* be transfinite; then $n1 + n2 = n1 \times n2 = \max(n1, n2)$. *End of aside.*

Be all that as it may, let's agree for the sake of the discussion that there are indeed an infinite number of transfinite cardinals. Following Cantor, we refer to these cardinals as "aleph numbers" (*aleph*, written $\aleph$, is the first letter of the Hebrew alphabet). So we can define an ascending sequence of such numbers—$\aleph_0$, $\aleph_1$, $\aleph_2$, etc.—as follows:

- $\aleph_0$ (*aleph zero*, also known as *aleph null*) is the smallest transfinite cardinal, and is the cardinality of the set of nonnegative integers (or of the set of natural numbers, if you prefer). Note that every countably infinite set has cardinality $\aleph_0$ by definition.

- For all $n > 0$, $\aleph_n$ is defined as the cardinality of the power set of any set of cardinality $\aleph_{n-1}$. (Loosely, $\aleph_n$ is 2^k, where k is $\aleph_{n-1}$. But note that to talk in terms of "$\aleph_{n-1}$" is pretty loose as well, since n might itself be some aleph number, in which case $n-1$ is indistinguishable from n.)

In particular, therefore, $\aleph_1$, the cardinality of the power set of the natural numbers, is 2^k where k is $\aleph_0$. ($\aleph_1$ is also the cardinality of the reals.) Now, Cantor tried, but failed, to prove what's called the *Continuum Hypothesis*: viz., that there's no set whose cardinality C lies strictly between $\aleph_0$ and $\aleph_1$, and hence that $\aleph_1$ is indeed the "next" transfinite cardinal after $\aleph_0$. However, it's now known—I'm deliberately simplifying here, somewhat—that the Continuum Hypothesis is independent of the axioms of conventional set theory, meaning that either it or its negation can be assumed to be true without leading to any inconsistencies.

Let me elaborate on this conclusion a little. Let propositions *P1* and *P2* be as follows:

P1: There exists a transfinite cardinal C such that $\aleph_0 < C < \aleph_1$.

P2: There doesn't exist a transfinite cardinal C such that $\aleph_0 < C < \aleph_1$.

Clearly, each of *P1* and *P2* is the negation of the other. Now, I'm not a logician, but I can certainly accept the idea that we can construct two distinct formal systems *F1* and *F2*, both founded on the same set of axioms except that one axiom *f1* of *F1* is replaced by its negation *f2* in *F2*. I can also accept the idea that both of those systems *F1* and *F2* might be internally consistent. But:

- If the real world—or the portion of the real world we're interested in, at any rate—is supposed to behave in accordance with some formal system, then at most one of *F1* and *F2* can be the formal system in question. In other words, albeit loosely once again: As far as that real world context is concerned, it clearly can't be the case that *P1* and *P2* both have the same truth value. Rather, if one is true, then the other must be false.

- But it also doesn't seem to make much sense to say in that same context that, e.g., *P1* is true and *P2* isn't, because the choice between them is, or appears to be, arbitrary. Is it really the case that the world could have "gone either way," as it were, on such a fundamental issue?

My own take on this matter—for what it's worth, which I freely admit is probably not much—is this: Since there doesn't seem to be any good reason for preferring *P1* over *P2* or the other way around, it's very tempting to suggest that both propositions are meaningless, and that the whole idea of transfinite numbers is nothing but a chimera.

For further discussion of the questions raised in this section, please see Appendix B to Part III of this essay.

APPENDIX A: ANOTHER EXAMPLE

In this appendix I offer another simple example to show why I think arguments based on diagonalization are or can be flawed. Let L_n be a list or sequence consisting of an initial sublist of the positive integers 1 to n ($n \geq 1$) in ascending order, followed by what I'll call the extension sublist, defined as follows:

- In the first position (i.e., position $n+1$ within L_n), the single positive integer $1+1$ (i.e., 2).

- In the next two positions (i.e., positions $n+2$ and $n+3$ within L_n), the two positive integers $2+1$ and $2+2$ (i.e., 3 and 4, respectively).

- In the next three positions (i.e., positions $n+4$, $n+5$, and $n+6$ within L_n), the three positive integers $3+1$, $3+2$, and $3+3$ (i.e., 4, 5, and 6, respectively).

And so on, up to and including:

- In the last n positions, the n positive integers $n+1$, $n+2$, ..., $n+(n-1)$, and $n+n$, respectively

Here's what L_n looks like for the first few values of n:

```
n               Initial sublist            Extension sublist

1               1                          2
2               1,2                        2,3,4
3               1,2,3                      2,3,4,4,5,6
4               1,2,3,4                    2,3,4,4,5,6,5,6,7,8
etc.
```

Observe now that for all n:

- The initial sublist contains n entries.

- The extension sublist contains $(n \times (n+1))/2$ entries, and for $n > 1$ that expression always evaluates to something greater than n; in fact, the larger the value of n, the more the extension sublist dwarfs the initial sublist.

- The extension sublist always contains some positive integers—in fact, exactly n of them, if we ignore duplicates—that don't appear in the initial sublist. But we don't conclude from this state of affairs that the set of positive integers in its entirety—which is clearly what the initial sublist consists of when n goes to infinity—is uncountable, or that it somehow misses some of the integers in the extension sublist. On the contrary, every positive integer in the extension sublist does eventually appear in the initial sublist as well.

- As n goes to infinity, so do the cardinalities of each of the two sublists. But we don't conclude from this state of affairs that the infinity that's the number of entries in the combined list is somehow "greater than" the infinity that's the number of entries in just the initial sublist. On the contrary, the initial sublist and the extension sublist, and hence the combined list, are all clearly countable, and that infinity is thus $\aleph_0$ in all three cases, by definition.

APPENDIX B: AND ANOTHER

After I'd completed the first draft of this part of the essay, I came across the paper "On Cantor's Important Proofs," by Wolfgang Mückenheim (*arXiv:math/0306200*). Among other things that paper contains the following striking (and I think persuasive) example. Suppose the given infinite list of infinite strings just happens to start like this—

```
0  0  0  0  . . .
1  0  0  0  . . .
1  1  0  0  . . .
1  1  1  0  . . .
. . .
```

—and so on; in other words, let the *i*th string in the initial square sublist consist of *i*−1 ones followed by nothing but zeros.[24] For all *n*, then, the choice algorithm (i.e., Cantor's diagonalization scheme) applied to the initial square sublist always produces "the next string," or in other words the first string in the remainder sublist.

APPENDIX C: CANTOR'S ORIGINAL PROOF

Here for the record is a translation, due to James R. Meyer, of Cantor's diagonal proof of the uncountability of the natural numbers:[25]

> In the paper [titled] *On a property of a set of all real algebraic numbers*[26] (Journal Math. Bd. 77, S. 258), there appeared, probably for the first time, a proof of the proposition that there is an infinite set of elements, which cannot be put into a one-one correspondence with the set of all finite whole numbers 1, 2, 3, ..., v, ..., or as I frequently state, the infinite set of elements does not have the magnitude of the number series 1, 2, 3, ..., v, ... From the proposition proved in [that paper] there follows another, that the set of all the real numbers in an arbitrary interval (a, b) cannot be arranged in the series:
>
> $$w_1, w_2, ..., w_v, ...$$
>
> However, there is a proof of this proposition that is much simpler, and which does not depend on a consideration of the irrational numbers. We let *m* and *n* [*w, surely, not n?—see below*] be two different characters, and consider a set *M* of elements:
>
> $$E = (x_1, x_2, ..., x_v, ...)$$
>
> where each element *E* is defined by infinitely many coordinates $x_1, x_2, ..., x_v, ...$ and where each of the coordinates is either *m* or *w*. Let *M* be the set of all such elements *E*. For example, we might have the following three elements of *M*:

---

[24] Of course, these first *n* strings all represent not just real but rational numbers ("rationals")—but so what? Cantor's diagonalization scheme certainly doesn't prohibit such a state of affairs. Nor of course does it prohibit a state of affairs in which the first *n* strings in the list happen to be the particular ones shown. *Note:* A rational number is a real number that's equal to the ratio of two integers (and then, of course, an irrational number is a real number that's not rational). Note that if the number of digits required for a precise representation of some number *x* (using any radix whatsoever) is finite, then that number *x* is certainly rational. Note also, however, that the converse of this statement is false (why, exactly?).

[25] Copyright © 2018 *www.jamesrmeyer.com*. My thanks to Meyer for permission to include his translation here. *Note:* Meyer prefaces his translation with the following: "The term *cardinality* for infinite sets was not in current usage at the time Cantor wrote this paper; he uses the term *Mächtigkeit*, which can have corresponding English meanings such as thickness, width, mightiness, potency, etc. I have used the term *magnitude* as a suitable translation." (Talking of terminology, incidentally, I note for the record that Meyer's translation frequently uses the term *series* when *sequence* would be more appropriate.)

[26] See Part II of this essay.

$$E_1 = (m, m, m, m, \ldots)$$
$$E_2 = (w, w, w, w, \ldots)$$
$$E_u = (m, w, m, w, \ldots)$$

I now assert that such a set M does not have the magnitude of the series 1, 2, 3, ..., v, ... This follows from the following proposition:

> If E_1, E_2, ..., E_v is any infinite series of elements of the set M, then there always exists an element E_0 of M, which cannot be the same element as any element E_v.

The proof is given by assuming that there are:

$$E_1 = (a_{1.1}, a_{1.2}, \ldots, a_{1.v}, \ldots)$$
$$E_2 = (a_{2.1}, a_{2.2}, \ldots, a_{2.v}, \ldots)$$
$$E_u = (a_{u.1}, a_{u.2}, \ldots, a_{u.v}, \ldots)$$
$$\ldots$$

where each $a_{u.v}$ is either m or w. Then there is a series b_1, b_2, ..., b_n, ... that can be defined so that b_v is also equal to m or w but is different from $a_{v.v}$. That is, if $a_{v.v} = m$, then $b_v = w$.

Consider the element:

$$E_0 = (b_1, b_2, b_3, \ldots)$$

of M. One sees straight away, that the equation:

$$E_0 = E_u$$

cannot be satisfied by any positive integer u, otherwise for that value of u and for all values of v, we would have:

$$b_v = a_{u.v}$$

and so we would in particular have:

$$b_u = a_{u.u}$$

which by the definition of b_v is impossible.

From this proposition it follows immediately that the set of all elements of M cannot be put into a sequence such as:

$$E_1, E_2, \ldots, E_v, \ldots$$

otherwise we would have a contradiction, that an element E_0 would be both an element of M, but also not an element of M.

This proof appears remarkable not only because of its great simplicity, but especially because the principles used in it can easily be extended to the general proposition that the magnitudes of well-defined sets have no maximum or, to the same effect, that for every given set L another set M can be generated which is greater than L.

APPENDIX D: A HARMLESS LITTLE ARGUMENT (?)

My own knowledge of logic derives in part from a certain popular account—a book, I mean—that I've had occasion to quote from, approvingly, in various previous writings. So when I came across a recent article by the author of the book in question concerning Cantor's diagonal proof, I read it eagerly. However, I soon found that the aim of that article wasn't to criticize Cantor's proof (not at all), but rather to debunk attempts to do so by other writers. In fact, the article characterizes Cantor's proof as "this harmless little argument," and asks, rhetorically, why so many people seem to be so angry with it.

Well, I'm not going to discuss that article in detail here; in particular, I most definitely don't want to give the impression that I think the author is wrong in his specific debunkings. I just want to note two things:

- As far as I can tell, the arguments the author uses in those debunkings don't apply to McGoveran's refutation as described in the main text of this part of the essay.

- The author concludes by stating categorically that *there's nothing wrong with Cantor's argument*—a claim that's very obviously at odds with my own position, of course.

Given that (as I've said) my previous experience with the author's work was positive, therefore, I decided to write to him with the specific aim of sounding him out on McGoveran's refutation. So I did. I began with a short introduction, and then I repeated, in slightly abbreviated form, the discussion of that refutation from pages 4-8 of this essay. And I finished up by saying:

If there's something logically wrong with the foregoing argument, I'd really like to know exactly what it is.

What follows is an edited version of the exchange that ensued. (When I say "edited," I mean I've made various changes to the original for cosmetic reasons, but never in such a way as to change the sense. I've also interspersed numerous additional responses by myself—responses, I mean, that weren't part of the original exchange and thus weren't seen by the other party.) For reasons that will sadly become clear only too quickly, I've referred to my correspondent throughout as "Critic."

Critic: You say: "If there's something logically wrong with the foregoing argument, I'd really like to know exactly what it is." OK, the first thing wrong is that you haven't explained what you mean by "the foregoing argument" (TFA). On my count you use this phrase four times, and I'm not sure you mean the same thing each time. But let me guess that "the foregoing argument" is [*McGoveran's refutation, which of course it is*]. If that's not right you can correct me. [*No correction needed.*]

There are some problems about what this argument TFA is. One is that the argument refers to "the list" but never says what list you are talking about. I think in fact it can be any list in order type ω which lists decimal expansions of distinct real numbers strictly between 0 and 1,[27] so I'll assume that.

Response: Well, I think my text is perfectly clear as to what "the list" refers to, but let me spell it out for the record. Actually there's one such list, or sequence, for every $n > 0$; for each such n, it's a list of all possible bit strings of length n, in arbitrary order. My text considers the case $n = 4$ in some detail, then goes on to extrapolate from that particular case to the general case (i.e., the case for all possible values of n).

I do want to quibble about that phrase "decimal expansions," though. First, my text makes it perfectly clear that we're talking about *binary* numbers (or representations of binary numbers, rather), not decimal.[28] Second, it talks very carefully about *bit strings*, not "expansions."

I also want to quibble about the phrase "any list [of] distinct real numbers" (or distinct bit strings, rather). To be specific, the list for any particular n isn't just any old list of any old bit strings (bit strings, that is, of the pertinent length n), it's a list of *all possible* bit strings (again, of the pertinent length n)—all 2^n of them, to be precise.

As for "order type ω": First, the symbol ω ("omega") denotes the first transfinite ordinal number. It's the ordinal counterpart to $\aleph_0$, the first transfinite cardinal number.[29] And to say a list L has order type ω is to say there's a one to one onto mapping between L and the natural numbers (in numerical order) that preserves ordering—i.e., for all i, the ith element of L corresponds under that mapping to the natural number i, and vice versa. For example, the list of all even positive integers in ascending order has order type ω. (By contrast, the list of those same integers in descending order doesn't.) Of course, if such a mapping does exist, then it follows immediately that L has the same number of elements as N, the set of natural numbers.

But I frankly don't know why we should have to drag in "order type ω" at all at this stage. McGoveran's refutation starts off, in effect, by considering *finite* lists and then asks what happens "at infinity" (i.e., when those lists approach infinite length). Talk of "ω" instead of just plain old "infinity" tends to suggest that the critic has already made up his mind that Cantor is right and the counterargument is wrong. I mean, it surely makes sense to talk about a *first*

---

[27] That phrase "strictly between 0 and 1" is a little odd. I take it to mean that 0 and 1 are explicitly excluded. But 0 is obviously not excluded (at least, not if we're talking about all possible bit strings of a certain length). And when the entries in the list become infinitely long, 1 isn't excluded either.

[28] As did Cantor himself, in effect (see Appendix C to this part of the essay); so why does the critic mention decimal at all?

[29] Just to remind you, the cardinals are the counting numbers (1, 2, 3, etc.); the ordinals denote position in a list or sequence (1st, 2nd, 3rd, etc.). And $\aleph_0$ is the cardinality of the set of all counting (or natural) numbers—which is to say, it's the cardinality of the set of all positive integers.

transfinite number only if you believe there are others—a second, a third, and so on—as well.

What's more, the idea that such things as ω and $\aleph_0$ might even exist never arose prior to Cantor's work; thus, those constructs certainly couldn't form any part of Cantor's original diagonal argument, [30] nor did they. Thus, to talk of either ω or $\aleph_0$ at this stage is, at best, hindsight; at worse, it's to risk getting into some kind of circular argument.

Finally: By definition, if the list of reals we're talking about really is of order type ω, then the reals in that list are countable by definition! So, given that the critic clearly believes the reals *aren't* countable, there must be something wrong with his suggestion that "the list" is "any list in order type ω which lists decimal expansions of distinct real numbers"—at least if the list in question is meant to be complete (i.e., to contain all possible reals), which as far as both Cantor and I are concerned it most definitely is.

Critic: Another problem is that you announce a lot of facts but it's not clear which if any of them is intended to be the conclusion of TFA. I'll assume that the conclusion is the conjunction of all of these facts.

Response: Actually I think my text is perfectly clear as to what the conclusion is, but again I'll state it explicitly for the record. Here it is: The diagonalization argument clearly fails in all finite cases; so why should we believe it's valid in the infinite case?

Critic: Now back to your question: "If there's something logically wrong with the foregoing argument, I'd really like to know exactly what it is." As far as I can see, there is nothing logically wrong with TFA as defined above. [*Good!*] However, TFA says nothing directly relevant to Cantor because it and its "conclusion" are entirely about finite arrays, and Cantor doesn't use finite arrays. You don't say how this is relevant to Cantor.

Response: I don't say how this is relevant to Cantor? Really? My exact words were:

> Diagonalization applies *by definition* only to the initial square sublist ... Since 2^n is always strictly greater than n, the remainder sublist is never empty ... The fact that diagonalization fails to produce an entry in the initial square sublist becomes less and less surprising, as it were, as n increases ... What's not clear to me is why, as Cantorians ... seem to believe, the foregoing argument should cease to be valid as n goes to infinity.

How this text could possibly be construed as not being "directly relevant to Cantor" is beyond me.

As for "Cantor not using finite arrays": I'm sorry, but this claim is absurd. I'll address the point in more detail in my next response.

---

[30] Nor of his interval argument, either (see Part II of this essay).

Critic: My guess is that you are assuming, for some property P that you don't spell out, if every finite number n has property P then ω must also have property P. You could be making a logical error here, or you may be making some false assumption about numbers—without seeing P spelled out I can't say.

Response: I don't spell out property P? Really? My exact words were:

> For any $n \geq 1$ the complete list of n-bit numbers can be divided into [an] *initial square* sublist, consisting of the first n entries (n bits wide and n entries deep) [and a] *remainder* sublist, consisting of the remaining $2^n - n$ entries (n bits wide and $2^n - n$ entries deep).

That's the property the critic refers to as property P.

Also, what I'm doing according to the critic's "guess" (viz., assuming that if every finite number has a certain property, then ω must have that property too) is *exactly* what Cantor's diagonal argument does too!—albeit without mentioning ω as such, of course. To spell the point out, Cantor shows that the antidiagonal differs from the nth string in the nth binary place, and then assumes that this property continues to hold when n becomes infinite. Why does the critic consider such a form of argument legitimate in Cantor's case but not in mine (or in McGoveran's, rather)? Why doesn't he accuse Cantor too of "making a logical error [or] making some false assumption about numbers"?

————— ◆◆◆◆◆ —————

So much for the critic's initial contribution to our exchange. I replied to it as follows:

> My apologies if my [previous attempt] wasn't sufficiently clear—I guess that's what happens if you take an excerpt from a longer essay (which is what I did) and edit it in an attempt to make it self-contained ... Let me [have another go]:
>
> Let n be an integer greater than zero.
>
> Let CL_n [*"Cantor list n"*] be a list (array) consisting of all possible bit strings of length n.
>
> CL_n is n bits wide and 2^n entries long.
>
> Let ISS_n be the first n strings of CL_n, and let RS_n be the remaining $2^n - n$ strings of CL_n.
>
> Observe that ISS_n is square (it's an $n \times n$ array).
>
> It therefore possesses a diagonal starting at top left. Note that the same couldn't validly be claimed if ISS_n weren't square.
>
> Let the diagonal of ISS_n starting at top left be the string d_n.
>
> Let the ones complement of d_n (i.e., the *antidiagonal* corresponding to d_n) be d_n'.

Let P_n be the proposition: $d_n' \notin ISS_n$ AND $d_n' \in RS_n$.

It's easy to show that P_n is true for all n. (That's what my previous email tried to do.)

My question is: Why, logically, should P_n cease to be true when n goes to infinity?

And my critic responded as follows:

Critic: Excellent. Your new version of the "foregoing argument" is close enough to what I guessed, and you've added some useful clarifications. I think we can agree that the "foregoing argument" in your new version is valid [*Good!*], though it does throw up a glitch that I'll come to below. So your question is just your final one:

Why, logically, should P_n cease to be true when n goes to infinity?

I think maybe you are begging a question by assuming logic has anything to do with this. It's a question of set theory. I'll ignore your word "logically."

Response: Several things I want to say here. First, "to beg the question" is usually understood to mean *assuming the very thing one is trying to prove*. I don't see how my question can possibly be taken as doing anything of the kind.

Second, the critic says "It's a question of set theory." What does he mean by "set theory"? It certainly can't be what's now known as *axiomatic* set theory, since that theory didn't exist when Cantor produced his diagonal argument; so a knowledge of that theory can't be required in order to understand or support (or indeed criticize) Cantor's argument. Thus, the significance of the critic's claim here is unclear at best.

Third, the critic says he'll ignore my word "logically." Well, if he means we should ignore *logic* in our debate—after all, he does explicitly say logic doesn't have "anything to do with" what we're talking about—then words fail me, and there's no point in our arguing any more.

Critic: The first step in answering your question is to clarify what you mean by "when n goes to infinity." No finite number literally "goes to infinity," so there is some shorthand here that needs to be spelled out.

Response: Of course I agree that "no finite number"—I'd prefer to say just "no number," because as far as I'm concerned there's no such thing as an infinite number—literally goes to infinity. No finite number "goes" anywhere, it just *is*. But phrases such as "n goes to infinity" or "n approaches infinity"—where, to be clear, n isn't a number as such but rather a parameter, or variable, that denotes a number—appear very commonly in the literature, and they serve as informal shorthand for some such statement as the following:

The set of values that are allowed as legal substitutions for n has no upper bound.

Clarification is always a good thing, of course, but in this particular case I really doubt whether the alleged lack of clarity in my original text makes that text ambiguous or hard to understand.

Critic: Most mathematicians would assume you have in mind some topological space S containing P_1, P_2, etc., as elements, and your question is about what happens "in the limit." This makes best sense if you are also assuming that S contains a limit element, say P_ω, where ω is the least transfinite cardinal. [*There's a muddle here which I'm not sure I can sort out. The least transfinite cardinal is $\aleph_0$—ω is the least transfinite ordinal, not cardinal. On the other hand, the 1, 2, etc. in P_1, P_2, etc. are cardinals, not ordinals. I mean, we do talk about P_1, P_2, etc., not P_{1st}, P_{2nd}, etc. However, I'll stay with the critic's use of ω as a subscript, in order to avoid a possibly irelevant debate.*]

Response: The critic's wording here seems to me just to be dressing up in fancy language what is after all a very simple idea. I would rather have said simply that we're dealing with an infinite set of propositions P_1, P_2, P_3, etc., that look like this:

```
P₁ :  d₁' ∉ ISS₁ AND d₁' ∈ RS₁
P₂ :  d₂' ∉ ISS₂ AND d₂' ∈ RS₂
P₃ :  d₃' ∉ ISS₃ AND d₃' ∈ RS₃
...  and so on, "forever"
```

Critic: What would P_ω say? One possibility is that it says

```
dω' ∉ ISSω AND dω' ∈ RSω
```

Response: "One possibility"? What other possibilities could there be?

Critic: So we need to work out what for example ISS_ω is. It's up to you to say what it is, since it was you who introduced the question of topology in your question.

Response: Gosh! I certainly wasn't aware that I'd "introduced the question of topology"—unless my critic is using that term now, not as previously (when he mentioned "topological spaces"), but merely to refer to the notion that ISS_n and RS_n have a certain ("topological") shape—to be specific, ISS_n is an $n \times n$ square and RS_n is an $n \times (2^n - n)$ rectangle.

Critic: It seems to me the most natural definition is that ISS_ω is a binary array[31] which has ω rows, each of order type ω. Since your [complete] "list" should

---

[31] Meaning, presumbly, simply that the element of the array at any given row and column intersection is either a 0 or a 1.

have order type ω (I'll come back to this), this array ISS_ω is the entire list, so there should be nothing left out, and RS_ω is empty.

Response: Talk about begging the question! It seems to me that the critic here is simply *assuming* that ISS_ω is the entire list and contains all possible bit strings. Or perhaps he's simply *defining* it that way (after all, he does use the word "definition"). Either way, it looks to me to be *exactly* a matter of assuming, or else asserting by fiat, the truth of the very thing he's trying to prove: in other words, begging the question.

It also seems to me that there's no earthly reason to believe any such thing (i.e., that ISS_ω contains all possible bit strings), since:

a. The analogous state of affairs most certainly doesn't hold for any finite n,

and moreover

b. The bigger n gets, the more that analogous state of affairs doesn't hold (if you see what I mean).

Let me elaborate for a moment on this latter point. As we know, for any given n the cardinality of the entire list CL_n is 2^n; the cardinality of the initial square sublist ISS_n is just n; and thus the cardinality of the remainder sublist RSn is $2^n - n$. Now let's focus on RS_n. It's easy to see that the cardinalities $|RS_n|$ for $n = 1, 2, 3, \ldots$ form a fast diverging sequence; in fact, given arbitrary $N > 0$, $|RS_n| > N$ for all $n \geq \log_2 N$.[32] As a consequence (to repeat something I said in the body of this part of the essay), as n increases RS_n very quickly dwarfs ISS_n, meaning that "almost all" of the values that appear in CL_n at all actually appear in RS_n and not in ISS_n. All of which, it seems to me, make the critic's assumption (or definition) that ISS_ω contains all possible bit strings unjustified in the extreme.

The critic goes on to say that the complete list has order type ω, and hence that ISS_ω is the entire list—there's nothing left out, and so RS_ω is empty. Well, the list in question (i.e., CL_ω) consists of ISS_ω followed by RS_ω. I agree it's infinite, if that's what the critic means when he says it has order type ω. But the fact that ISS_ω and the complete list CL_ω both have order type ω isn't sufficient to prove they're identical (i.e., to prove that ISS_ω is the entire list and RS_ω is empty accordingly).[33] After all, it's the very essence—indeed, it's the definition—of an infinite set that it can be put into one to one correspondence with some proper subset of itself.[34] So the set consisting of all reals in CL_ω and the proper subset consisting of all reals in ISS_ω can perfectly well both have order type ω without there being any implication that ISS_ω has to be the whole of CL_ω, and hence that

---

[32] If we replaced that $N > 0$ by $N > 1$, we could replace that $n \geq \log_2 N$ by the stronger inequality $n > \log_2 N$. Here for interest are the first few values of $|RS_n|$: 1, 2, 5, 12, 27, 58, 121, ...

[33] For example, the set of all integers and the set of all even integers both have order type ω, but they're obviously not identical.

[34] And I see no reason in general why such a correspondence shouldn't be order preserving, if the critic or someone thinks that's important—though I note in passing that Cantor didn't seem to think it was.

the subset isn't a proper subset after all. So I'd say that the critic's claim—viz., that ISS_ω is the entire list—remains unproven, at the very least.

Critic: Now we can see what P_ω says. First,

$$d_\omega' \notin ISS_\omega$$

This is true directly by Cantor's diagonal argument, and in fact this is Cantor's theorem.

Response: Yes, it's true "by Cantor's diagonal argument." But so what? It doesn't follow—at least, it doesn't follow from this fact alone—that d_ω' doesn't appear elsewhere in the complete list CL_ω (i.e., in the remainder sublist RS_ω).

As for "in fact this is Cantor's theorem": So the critic apparently agrees that "Cantor's theorem" shows only that the antidiagonal doesn't appear in the initial square sublist!—*not* that it doesn't appear in the list at all. Well, good, I suppose.

PS (a quibble): The term "Cantor's theorem" as used by the critic here refers, of course, to Cantor's diagonal proof. By contrast, the term "Cantor's Theorem"—note the uppercase T—is generally understood to refer to Cantor's result to the effect that the cardinality of any given set is strictly less than that of the corresponding power set. This latter result is obviously not what we're talking about here; nevertheless, I think my critic should have been a little more careful in his wording—especially since he complained earlier about an alleged lack of clarity on my own part.

Critic: [Second,]

$$d_\omega' \in RS_\omega$$

This is false because RS_ω is empty. So we showed that the P_n for finite n are true, and separately we showed that P_ω is false. Does this answer your question?

Response: The critic claims to have shown that (a) RS_ω is empty and so (b) P_ω is false. I don't believe he's done either of these things, and I've explained why I don't. Moreover, I've also given what seems to me to be a pretty strong argument for believing RS_ω is *not* empty. From all of which I conclude (this text is repeated, albeit very slightly edited, from the body of this part of the essay):

> Cantor claimed to have proved that the reals are uncountable; what's more, his proof seems to be very widely accepted. However, [*McGoveran's argument among others*] demonstrates rather clearly that the proof in question is flawed, and in fact no proof at all. Note carefully, however, that we can't reasonably conclude from this state of affairs that the reals are countable—all we can conclude is that the diagonal argument doesn't prove they aren't.

And no, the critic hasn't answered my question.

Critic: Now for the glitch. You say "Let CL_n be a list (array) consisting of all possible bit strings of length n." That's fine. But then when you come to your question, you may be presupposing that there is a corresponding proposition P_ω which talks about the "square" ISS_ω which consists of the first ω rows of CL_ω, i.e., the entire list of reals.

Response: No! The proposition P_ω talks about the whole of CL_ω, not just ISS_ω. CL_ω consists of ISS_ω followed by RS_ω. Now, I agree that ISS_ω is square. I also agree it consists of the first ω rows of CL_ω.[35] But, to repeat myself, I don't agree it contains the entire list of reals, because I don't believe RS_ω is empty; a fortiori, therefore, I obviously don't believe it's been *shown* to be empty.

As a matter of fact, if ISS_ω did contain the entire list of reals, it would follow immediately, from the very definition of ω, that the reals are countable!— which as I said before the critic clearly doesn't believe, so he seems to be contradicting himself here.

Critic: Your remark about "all possible bit strings of length n" suggests that CL_ω should contain all real numbers.

Response: Correct.

Critic: But by Cantor's theorem this is impossible, so you have landed yourself in a self-contradictory situation.

Response: Begging the question again! I'm arguing that Cantor's theorem to the effect that the reals are uncountable isn't in fact a theorem at all: specifically, that it doesn't prove what it claims to prove. So the critic can hardly appeal to that very theorem in his attempt to discredit my argument. Thus, I categorically deny the accusation that I've "landed myself in a self-contradictory situation."

Critic: But in fact this [*i.e., landing yourself in a self-contradictory situation*] is not necessary; you can simply remove the words "all possible" in your description of CL_n. Cantor's argument doesn't need it.

Response: I don't understand these remarks at all. For any given n, the *whole point* about CL_n is that it consists of all possible bit strings of length n (and, of course, that such remains the case as n goes to infinity). Removing the words "all possible" would leave nothing but nonsense—certainly not a proof, not even of any kind.

As for "Cantor's argument doesn't need it": To the extent that I understand what the critic intends by that claim, I would have said that (a) Cantor's argument is supposed to be a proof (or "proof") by contradiction, and hence that (b) it most certainly does need that hypothesis (viz., that CL_n contains all possible bit strings of length n, and that such remains the case as n goes to

---

[35] "The first ω rows"? Hardly a very precise characterization, of course (especially since ω is an ordinal, not a cardinal), but they're the critic's own words.

infinity). Specifically, it needs it as the initial hypothesis that leads to the alleged contradiction.

Concluding Remarks

If believers in Cantor's diagonal proof can't do better than the critic cited above in defense of their position, then I feel bound to say their position does seem to be an extraordinarily weak one—which makes it all the more curious that it seems to be almost universally accepted by mainstream mathematicians. How can this latter state of affairs (the widespread acceptance of Cantor's proof, I mean) be explained? Well, it's only a guess, of course, but my suspicion is that what we have here is a textbook example of "the law of increasing returns"—or something very like that law, at any rate. What I mean by this remark is the following:

> If N people believe X, then that fact makes it easier for the $(N+1)$st person to believe X as well.

Or to put it another way:

> If N people believe X, then that fact makes it harder for the $(N+1)$st person to argue against X.

Moreover, once N has reached some critical threshold, X becomes generally accepted—at which point few people have either the time or the inclination to go back and examine X again (from first principles, as it were) and consider whether it really stands up to careful analysis.

Part II:

Cantor's Interval Argument

Lucid intervals and happy pauses

—Francis Bacon:
History of King Henry VII (1622)

In Part I of this essay I explained why I'm skeptical about Cantor's diagonal proof of the uncountability of the reals. Again let me repeat the following remarks, since they capture the essence of the argument thus far:

> Cantor claimed to have proved [*i.e., using his diagonal scheme*] that the reals are uncountable; what's more, his proof seems to be widely accepted. However, I believe the arguments [*the ones presented in Part I of this essay, that is*] demonstrate rather clearly that the proof in question is flawed, and in fact no proof at all. Of course, we can't reasonably conclude from this state of affairs that the reals are countable—all we can conclude is that the diagonal argument doesn't prove they aren't.

Actually, however, that diagonal proof, which was published in 1891, was Cantor's second attack on the problem—he'd already published another less well known proof some years previously, in 1874, in a paper titled "On a Property of the Collection of All Real Algebraic Numbers" ("*Über eine Eigenschaft des Inbegriffes aller reellen algebraischen Zahlen*").[1] The article "Cantor's 1874 Proof of Non-Denumerability," by James R. Meyer,[2] includes an English translation of that earlier proof (and that translation is reproduced as Appendix A to the present part of this essay). Here's a precise statement of what the proof in question allegedly proves (I say "allegedly" because I believe that, like his diagonal proof, Cantor's interval proof is flawed, and hence that what it claims to prove needs to be regarded with some circumspection):

> Let *I* be a real interval; then there's no real sequence *X* that contains every real in *I*.

---

[1] *http://www.digizeitschriften.de/main/dms/img/?PPN=GDZPPN002155583*. To elaborate briefly: A real number is *algebraic* if it's a solution to a polynomial equation with integer coefficients, and *transcendental* otherwise (thus, e.g., $\sqrt{2}$, being a solution to the equation $x^2 - 2 = 0$, is algebraic, while π is transcendental), and Cantor proves in his paper that the set of real algebraic numbers is countable. That's the property referred to in the paper's title. His proof, to be discussed in detail in what follows, that by contrast the reals aren't countable occurs later in the same paper. (As a matter of fact, there's reason to believe he "buried" this latter proof deliberately, in an attempt to avoid the attacks he felt—with reason, as it turned out—he'd be subjected to if it got a lot of attention.)

[2] *https://www.jamesrmeyer.com/pdfs/Cantor-1874-Proof-of-Non-Denumerability.pdf*.

Clearly, the precise meaning of this result depends on what's meant by the terms *real interval* and *real sequence*, and I'll give definitions of those terms, as well as a number of others, in the section "Definitions" below. Here let me just note in the interest of accuracy that—despite the title of Meyer's article—Cantor's paper actually *doesn't* contain an explicit proof that the reals are uncountable. However, what it does prove (or, rather, what it claims to prove, viz., that there's no X that contains every real in I) does have the uncountability of the reals as a corollary. In other words, if that proof were valid, it would indeed follow that the reals were uncountable (or "nondenumerable"), and for that reason Cantor's paper is usually thought of as having proved this latter proposition. See the section "What Cantor's Result Implies," later.

The aim of what follows, then, is to examine this earlier argument of Cantor's—which I'll refer to for present purposes as *Cantor's interval proof*—in detail. I'll begin with some definitions, then give my own explanation of the proof in question, followed by examples. Then I'll raise and discuss a number of what seem to me to be pertinent questions.

DEFINITIONS

Note: As you might have noticed, I'm making no attempt in this essay to give a formal definition for the term "real number" ("real" for short).[3] Informally, however, a real can be thought of as a point on the real number line—also not defined here!—and a real interval can be thought of as a section, or segment, of that line. Also, I remind you from Part I of this essay that I'm ignoring reals less than zero (barring explicit statements to the contrary, of course).

Definition (natural number): A positive integer.
Note: As mentioned in Part I of this essay, some writers additionally consider zero to be a natural number; the literature isn't consistent on this point, but most writers, and indeed common usage, do exclude the zero case, and in what follows I'll do the same.

Definition (countable, of a set): Capable of being put into strict one to one correspondence with the set of natural numbers or some proper subset thereof. Also known as *numerable*, *enumerable*, or *denumerable*.
Note: The significance of that phrase "or some proper subset thereof" is that it allows the term *countable* to be used of finite sets as well as infinite ones. The term *countably infinite* is sometimes used (and was used in Part I of this essay) to describe a set that's countable but infinite. The set of natural numbers itself—sometimes denoted N—is an example of a countably infinite set. Also, the term *countable* is sometimes used in connection with various other kinds of "collections," including sequences in particular (see the definition immediately following).

---

[3] However, if you'd like to see a precise formal definition, together with a very clear explanatory discussion, then I recommend G. H. Hardy's classic text *A Course of Pure Mathematics* (10th edition, Cambridge University Press, 1952).

Definition (real sequence): A countable collection—call it X—of real numbers $x_1, x_2, ..., x_n, ...$ for which order matters (i.e., x_1 is the first real in X, x_2 is the second real in X, ..., and more generally x_n is the nth real in X). The reals x in X are the elements of X and are said to be contained in X (written $x \in X$). If the number of elements contained in X is finite then X is a finite sequence; otherwise it's an infinite sequence (but is still countable, by definition). Duplicate elements are allowed, barring explicit statements to the contrary.

Definition (real interval): Let a and b be reals such that $a \leq b$. Then the real interval $(a...b)$—call it I—is the set of reals x such that $a < x < b$. The reals x in I are the elements of I and are said to be contained in I (written $x \in I$), and a and b are the boundary values of I. Observe that the boundary values aren't themselves elements of I—i.e., a and b aren't contained in I (written $a \notin I$, $b \notin I$). Observe further that if $a = b$, then I is empty; by contrast, if $a < b$, then I contains an infinite number of elements, because the set of reals in their entirety—in other words, the real number line—is "everywhere dense" (see Appendix D to the present part of this essay).

Points arising from this last definition:

■ Observe first that what the definition refers to as a real interval could more specifically—or more precisely, rather—be called an *open* real interval (see further discussion below).

■ All intervals to be discussed in this essay will be real intervals specifically. In other words, all elements of all intervals discussed will be real numbers specifically, and the unqualified term *interval* will be taken to mean a real interval specifically (barring explicit statements to the contrary in both cases, of course). However, I note for the record that other kinds of intervals and elements are certainly possible; for example, the elements that make up a temporal interval are points in time.[4]

■ Observe next that I define the real interval $(a...b)$ as having "boundary values" a and b. Now, some writers would refer to those values a and b simply as *bounds*; however, I find it useful to distinguish between bounds as a general concept, on the one hand, and boundary values as a special case of that more general concept on the other. By way of example, consider the real interval $I = (1.0...5.0)$. The boundary values a and b for that interval are of course 1.0 and 5.0, respectively. By contrast, any real a' such that $a' \leq 1.0$ is a bound (actually a *lower* bound) for I, and the boundary value $a = 1.0$ is the greatest of those lower bounds (GLB for short). Similarly, any real b' such that $b' \geq 5.0$ is also a bound (an *upper*

---

[4] For a much more extensive discussion of intervals in general and temporal intervals in particular, see the book *Time and Relational Theory: Temporal Databases in the Relational Model and SQL*, by C. J. Date, Hugh Darwen, and Nikos A. Lorentzos (Morgan Kaufmann, 2014).

bound) for I, and the boundary value $b = 5.0$ is the least of those upper bounds (LUB for short).

- The definition as stated is tailored to the purposes of this essay (or the present part of this essay, at any rate). However, let me note for the record that, even if we limit our attention to intervals whose contained elements are real numbers, the intervals in question can still be of several different kinds, and it's convenient to explain those various different kinds here:[5]

    1. The *open / open* interval *(a...b)*—so called because it's "open" at both ends, meaning the boundary values a and b aren't contained in the interval—is the set of reals x such that $a < x < b$. More usually known just as an *open* interval for short.

    2. The *closed / closed* interval *[a...b]*—so called because it's "closed" at both ends, meaning the boundary values a and b are contained in the interval—is the set of reals x such that $a \le x \le b$. More usually known just as a *closed* interval for short. Observe the notational use of brackets instead of parentheses.

    3. The *open / closed* interval *(a...b]* is the set of reals x such that $a < x \le b$ (so b is contained in the interval but a isn't). Usually known as a *half open* or *half closed* interval—more specifically, *open below and closed above*.

    4. The *closed / open* interval *[a...b)* is the set of reals x such that $a \le x < b$ (so a is contained in the interval but b isn't). Usually known as a *half closed* or *half open* interval—more specifically, *closed below and open above*.

All of that being said, please note again that from this point forward the term *interval*, unqualified, will be taken in this part of the essay to mean an open / open interval specifically (once again barring explicit statements to the contrary, of course).

Finally, I also need to define what it means for one interval to be strongly included in another:

Definition (strong inclusion): Interval $I_j = (a_j...b_j)$ is strongly included in interval $I_i = (a_i...b_i)$ if and only if $a_j > a_i$ and $b_j < b_i$ (i.e., every element of I_j is an element of I_i, and neither of the boundary values of I_j is a boundary value for I_i).

_____

[5] Several of the terms introduced in the discussion that follows are nonstandard—but unlike some of the more standard terms, at least they're systematic.

CANTOR'S INTERVAL PROOF

Note: As I've said, I believe Cantor's interval proof is flawed. In order to be able to demonstrate the flaw, however, I obviously have to present the proof as such first. For the time being, therefore, I'm going to have to suspend disbelief, as it were, and do my best to explicate that proof as carefully as I can. That's the purpose of the present section.

I'll begin by stating, using the formal style or structure I learned at school for such things, exactly what Cantor's interval proof purportedly proves.

Given: A nonempty interval $I_0 = (a_0...b_0)$ and an infinite sequence $X = x_1, x_2, ..., x_n, ...$, where $a_0, b_0, x_1, x_2, ..., x_n, ...$ are reals and we assume without loss of generality that X contains no duplicates (i.e., the x's are all distinct from one another). *Note:* As you can see, I choose to refer to the given interval in this section (also in most subsequent sections) not as I but as I_0.

To prove: There exists a real $y \in I_0$ such that $y \notin X$. *Note:* This result would obviously hold if X were finite, which is why we assume it isn't.

Construction: An ordered collection of intervals $I_1 = (a_1...b_1)$, $I_2 = (a_2...b_2)$, ..., $I_n = (a_n...b_n)$, ..., such that, for all $i \geq 0$, I_{i+1} is strongly included in I_i (for further specifics, see below).

Proof: First of all, note that I_0 and X both contain an infinite number of elements—I_0 by definition (because it's a nonempty real interval), and X because it's so stipulated as one of the givens.

    If I_0 and X have no elements in common—e.g., if I_0 is $(1.0...3.0)$ and X is all reals greater than 5.0—then what's to be proved is trivially the case. So assume from this point forward that I_0 and X do have at least one element x in common.

    Next, if I_0 and X have just one element x in common—e.g., if I_0 is $(1.0...3.0)$ and X is 2.0 followed by all reals greater than 5.0—then no element of I_0 apart from x appears in X, and again what's to be proved is trivially the case. So assume from this point forward that I_0 and X do have at least two distinct elements in common.

Aside: Actually the previous two paragraphs could be deleted without affecting the validity of the proof overall, because they're effectively subsumed by subsequent arguments. I include them only because they immediately remove certain special cases from further consideration, thereby (I believe) making the general proof a little easier to follow. *End of aside.*

Now let x_i and x_j be the first two distinct elements of X that are contained in I_0 ("first two" meaning, here and throughout the proof, "first two with respect to the ordering of elements of X," of course). Without loss of generality assume $x_i < x_j$ (so $a_0 < x_i < x_j < b_0$). Let I_1 be the interval $(a_1...b_1)$, where $a_1 = x_i$ and $b_1 = x_j$. Observe that I_1 is strongly included in I_0.

Now let x_k and x_l be the first two distinct elements of X that are contained in I_1. Without loss of generality assume $x_k < x_l$ (so $a_1 < x_k < x_l < b_1$). Let I_2 be the interval $(a_2...b_2)$, where $a_2 = x_k$ and $b_2 = x_l$. Observe that I_2 is strongly included in I_1.

Continuing in this fashion, let intervals $I_3, I_4, ..., I_n, ...$ be defined analogously. Observe that:

- The set of intervals so constructed is countable by definition.

- For all p ($p = 1, 2, ...$) I_p is strongly included in I_{p-1}.

- For all p ($p = 1, 2, ...$) the boundary values a_p and b_p of interval I_p are elements of X (also of $I_0, I_1, ..., I_{p-1}$) such that $a_p < b_p$.

- The numbers $a_0, a_1, a_2, ...$ are strictly monotonically increasing in magnitude (i.e., $a_0 < a_1 < a_2 < ...$) and the numbers $b_0, b_1, b_2, ...$ are strictly monotonically decreasing in magnitude (i.e., $b_0 > b_1 > b_2 > ...$).

Now there are two possibilities:

1. The total number of intervals so constructed is finite. In this case, let the last interval constructed be $I_n = (a_n...b_n)$. Observe that I_n contains at most one element x of X (for otherwise it would be possible to define a further interval I_{n+1}, strongly included in I_n, with both boundary values being elements of X). But I_n contains an infinite number of elements, by definition; thus, every element $y \in I_n$ apart from x (if it exists) is an element of I_n, and hence of I_0, that doesn't appear in X.

2. The total number of intervals so constructed is infinite. In this case, the sequence $a_0, a_1, a_2, ...$ has a limiting value a_∞, because the elements of that sequence are of finite magnitude and strictly monotonically increasing; likewise, the sequence $b_0, b_1, b_2, ...$ also has a limiting value b_∞, because the elements of that sequence are of finite magnitude and strictly monotonically decreasing. (These claims are valid because there exist theorems to the effect that (a) if a strictly monotonically increasing sequence has an upper bound, then it has a *least* upper bound (LUB), and

the elements of that sequence have a limiting value that's precisely that LUB; similarly, (b) if a strictly monotonically decreasing sequence has a lower bound, then it has a *greatest* lower bound (GLB), and the elements of that sequence have a limiting value that's precisely that GLB.)[6] Clearly $a_\infty \le b_\infty$. Now there are two subcases to consider:

i. If $a_\infty < b_\infty$, then the argument of paragraph 1 above—opening sentence excluded, of course—applies, mutatis mutandis, with a_∞ and b_∞ playing the roles of a_n and b_n, respectively.

ii. If $a_\infty = b_\infty$ (in which case the corresponding interval I_∞ is empty), then $y = a_\infty = b_\infty$ is itself an element of I_0 not appearing in X. For if it did appear in X, then we would have $y = x_n$ for some n. But as the lemma below implies, $x_n \notin (a_n...b_n)$ for all n; so y ($= x_n = a_\infty = b_\infty) \notin I_n$. But a_∞ and b_∞ are certainly elements of every interval that precedes the interval $I_\infty = (a_\infty...b_\infty)$ in that ordered collection of constructed intervals, and so they're both elements of I_n in particular (i.e., $y \in I_n$). So we have a contradiction (viz., $y \in I_n$ and $y \notin I_n$). It follows that $y = a_\infty = b_\infty$ is an element of I_0 that doesn't appear in X.

In all three cases (1, 2i, and 2ii), therefore, $y \in I_0$ and $y \notin X$. QED.

Lemma: For all $n > 0$ and for all i ($i = 1, 2, ..., 2n$), $x_i \notin I_n$ (where $I_n = (a_n...b_n)$).

<u>Basis</u> (show that $x_1 \notin I_1$ and $x_2 \notin I_1$): Consider $I_1 = (a_1...b_1)$, the first constructed interval, and x_1, the first element of X. Clearly there are three possibilities:

1. $x_1 \le a_0$. In this case, $x_1 \notin I_0$ and so $x_1 \notin I_1$, a fortiori.

2. $x_1 \ge b_0$. In this case again, $x_1 \notin I_0$ and so $x_1 \notin I_1$, a fortiori.

3. $a_0 < x_1 < b_0$. In this case, x_1 is, necessarily and by definition, the first element of X to be contained in I_0. By definition of I_1, therefore, x_1 is either a_1 or b_1; thus, again, $x_1 \notin I_1$.

Now consider that same interval $I_1 = (a_1...b_1)$ once more and x_2, the second element of X. Again there are three possibilities:

1. $x_2 \le a_0$. In this case, $x_2 \notin I_0$ and so $x_2 \notin I_1$, a fortiori.

2. $x_2 \ge b_0$. In this case again, $x_2 \notin I_0$ and so $x_2 \notin I_1$, a fortiori.

---

[6] The concept of a *limiting value*, or just a *limit* for short, is defined in Appendix C to this part of the essay. Here for simplicity I'm assuming the meaning is intuitively obvious. I observe for the record, though, that unlike the boundary values $a_1, a_2, ..., b_1, b_2, ...,$ those limiting values a_∞ and b_∞ aren't obtained by that recursive construction process but are merely asserted—of course correctly—to exist. Note, therefore, that there's no guarantee that a_∞ and b_∞ actually appear as elements of X. (Nor of course is there any guarantee that they don't.)

3. $a_0 < x_2 < b_0$. In this case, x_2 is, necessarily and by definition, either the first or the second element of X to be contained in I_0 (it's the second if x_1 is the first—see above—otherwise it's the first). By definition of I_1, therefore, x_2 is either a_1 or b_1; thus, again, $x_2 \notin I_1$.

<u>Inductive step</u> (assume that for some $n \geq 1$ and for all i in the range 1 to $2n$ inclusive, $x_i \notin I_n$, and show that it follows that $x_{2n+1} \notin I_{n+1}$ and $x_{2n+2} \notin I_{n+1}$):

Consider the $(n+1)$st interval $I_{n+1} = (a_{n+1}...b_{n+1})$ and x_{2n+1}, the $(2n+1)$st element of X. Once again there are three possibilities:

1. $x_{2n+1} \leq a_n$. In this case, $x_{2n+1} \notin I_n$ and so $x_{2n+1} \notin I_{n+1}$, a fortiori.

2. $x_{2n+1} \geq b_n$. In this case again, $x_{2n+1} \notin I_n$ and so $x_{2n+1} \notin I_{n+1}$, a fortiori.

3. $a_n < x_{2n+1} < b_n$. In this case, by our inductive step assumption, x_{2n+1} is the first element of X to be contained in I_n. By definition of I_{n+1}, therefore, x_{2n+1} is either a_{n+1} or b_{n+1}; thus, again, $x_{2n+1} \notin I_{n+1}$.

Now consider that same $(n+1)$st interval $I_{n+1} = (a_{n+1}...b_{n+1})$ once again and x_{2n+2}, the $(2n+2)$nd element of X. Yet again there are three possibilities:

1. $x_{2n+2} \leq a_n$. In this case, $x_{2n+2} \notin I_n$ and so $x_{2n+2} \notin I_{n+1}$, a fortiori.

2. $x_{2n+2} \geq b_n$. In this case again, $x_{2n+2} \notin I_n$ and so $x_{2n+2} \notin I_{n+1}$, a fortiori.

3. $a_n < x_{2n+2} < b_n$. In this case, by our inductive step assumption, x_{2n+2} is either the first or the second element of X to be contained in I_n (it's the second if x_{2n+1} is the first—see above—otherwise it's the first). By definition of I_{n+1}, therefore, x_{2n+2} is either a_{n+1} or b_{n+1}; thus, again, $x_{2n+2} \notin I_{n+1}$.

<u>Conclusion</u>: For all $n > 0$ and for all i ($i = 1, 2, ..., 2n$), $x_i \notin I_n$. QED.

The foregoing lemma has a number of significant corollaries:

1. The conclusion can be restated thus: If the ordering of X is thought of as proceeding from left to right with x_1 at the far left, then for all $n > 0$ and for all i ($1 \leq i \leq 2n$), x_i appears in X strictly to the left of a_n.

2. Consider the interval $I_n = (a_n...b_n)$. Then a_n is x_p for some $p \geq 2n-1$. *Note:* Actually p will be equal to $2n-1$ if and only if none of $x_1, x_2, ..., x_{2n}$ is "out of range," as it were—in other words, if and only if I_1 is $(x_1...x_2)$, I_2 is $(x_3...x_4)$, ..., and I_n is $(x_{2n-1}... x_{2n})$.

3. Likewise, b_n is x_q for some $q \geq p$ (and so $q \geq 2n-1$ also). *Note:* Actually q will be equal to p only in the case of the limiting interval I_∞, and then only

if we're talking about what the proof refers to as Case 2ii. Apart from that special case, q will always be strictly greater than p, and hence greater than or equal to $2n$.

4. Note in particular that (a) for all $n > 1$, x_n appears strictly to the left of a_n, and (b) for all $n > 0$, $x_n \notin I_n$. This latter point (b) is the specific point I appealed to in Case 2ii in the main proof earlier.

5. (*Important!*) Note that for all $n > 0$, if x_m is an element of X that appears in I_n, then $m > q$, and hence $m > 2n$; in other words, x_m appears in X strictly to the right of b_n.

The following diagram sketches the general situation at this point:

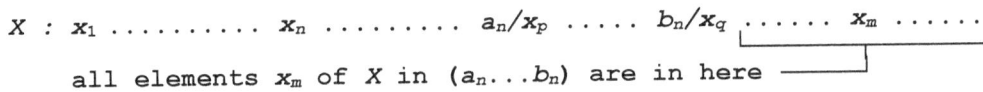

```
X : x₁ ........... xₙ .......... aₙ/xₚ ..... bₙ/xq ...... xₘ .....
                                                    ⌐_____⌐
     all elements xₘ of X in (aₙ...bₙ) are in here ————⌐
```

I'll come back to these points near the end of the section "Some Questions," later.

EXAMPLES

You might have found the arguments in the previous section a little abstract and hence somewhat hard to follow—so let's look at a couple of concrete examples, both of them based on one originally constructed by Robert White. First let the given sequence X be as follows:

$$
\begin{aligned}
x_1 &= 1.9 \\
x_2 &= 2.1 \\
x_3 &= 1.99 \\
x_4 &= 2.01 \\
x_5 &= 1.999 \\
x_6 &= 2.0 \\
x_7 &= 2.001 \\
x_8 &= 1.9999 \\
x_9 &= 2.0001 \\
x_{10} &= 1.99999 \\
x_{11} &= 2.00001 \\
x_{12} &= 1.999999
\end{aligned}
$$

And so on. You can think of this sequence—but, I stress, only informally; I'm *not saying the sequence is formally defined in this fashion*—as being constructed as follows: Start with two infinite subsequences, one of strictly increasing values of the form 1.9, 1.99, 1.999, 1.9999, etc., the other of strictly decreasing values of the form 2.1, 2.01, 2.001, 2.0001, etc.; then merge those two subsequences, taking elements alternately from the first subsequence and then the second; then inject a solitary 2.0 between, say, 1.999 and 2.001 (in other words, as the sixth element), that position in the overall sequence being chosen arbitrarily.

Now let a_0 and b_0 be 1.9 and 2.1, respectively; i.e., let the given interval I_0 be (1.9...2.1). Observe that neither x_1 nor x_2 is an element of this interval. Thus we have:

$$I_1 = (1.99 \quad\; ... 2.01 \quad) - \text{i.e., } (a_1...b_1) \text{ is } (x_3... x_4)$$
$$I_2 = (1.999 \;\; ... 2.0 \quad\;\;) - \text{i.e., } (a_2...b_2) \text{ is } (x_5... x_6)$$
$$I_3 = (1.9999 ... 1.99999) - \text{i.e., } (a_3...b_3) \text{ is } (x_8...x_{10})$$

I_3 is the last interval to be constructed in this example, because $b_3 = 1.99999$, and there's no x in the sequence X that (a) is strictly less than that value 1.99999 and (b) hasn't already been used as a boundary value for some previous interval. So every y such that $1.9999 < y < 1.99999$ (e.g., $y = 1.99995$) is an element of I_0 that's not an element of X.

For our second example, suppose the given sequence X is as before, except that the element 2.0 is removed. Then we have:

$$I_1 = (1.99 \quad\;\; ... 2.01 \quad\;) - \text{i.e., } (a_1...b_1) \text{ is } (x_3... x_4)$$
$$I_2 = (1.999 \quad ... 2.001 \quad) - \text{i.e., } (a_2...b_2) \text{ is } (x_5... x_6)$$
$$I_3 = (1.9999 \;\; ... 2.0001 \;\;) - \text{i.e., } (a_3...b_3) \text{ is } (x_7... x_8)$$
$$I_4 = (1.99999 ... 2.00001) - \text{i.e., } (a_4...b_4) \text{ is } (x_9...x_{10}) ... \text{etc.}$$

This time the process of constructing new intervals never terminates. Note, however, that the a's and b's now both have the limiting value 2.0, and 2.0 is an element of every constructed interval[7] (as well as of the original interval I_0); thus $a_\infty = b_\infty = 2.0$ is an element of I_0 that doesn't appear in X.

WHAT CANTOR'S RESULT IMPLIES

Here again is what Cantor's interval proof purportedly proves:

> Let I be a real interval; then there's no real sequence X—no infinite real sequence X in particular—that contains every real in I.[8]

For brevity, I'll refer to this statement as "Cantor's Interval Theorem." Observe now that what this theorem *doesn't* say—at least, not explicitly—is that the reals in I are uncountable. What's more, I don't think it's immediately obvious that this latter is a logical consequence of what it does say, either. After all, consider the following argument:

---

[7] With the sole exception of the interval $I_\infty = (a_\infty...b_\infty)$, if you think that one's "constructed."

[8] From the remainder of this part of the essay I'll refer to the given interval sometimes (as here) just as I, sometimes more specifically as I_0.

- Let X and I be some arbitrary real sequence and some arbitrary real interval, respectively.

- Let X and I both be of infinite cardinality. (Of course, I will be of infinite cardinality by definition, unless it's empty.)

- Given such an X and I, it's very unlikely, in general, that they'll both contain exactly the same set of elements.

- Thus, the idea that I might contain a real y that doesn't appear in X is hardly surprising. (Of course, it might also be the case that X contains a real y' that doesn't appear in I. Indeed, it would be surprising if it didn't.)

- So you might be thinking that we could (a) just extend X to contain that "missing value" y, and then (b) repeat the process for all such "missing values." Then eventually (c) it would no longer be the case that I contains a value not in X.

And indeed the foregoing argument would be valid *so long as it can be guaranteed that the set of "all such missing values" is countable.*[9] But of course no such guarantee can be made, and the argument is thus not valid after all. By contrast, the following argument is valid:

- Suppose the reals in I are countable; i.e., suppose there exists a one to one onto mapping, M say, from the reals in I to the natural numbers.[10]

- Consider the ordered pair $p_i = <r_i,n_i>$, where r_i is a real in I and n_i is the natural number corresponding to r_i under M.

- Since M is one to one and onto, no two distinct such pairs can contain the same r or the same n; that is, if $j \neq k$, then $r_j \neq r_k$ and $n_j \neq n_k$.

- Let S be the set of all such ordered pairs p_i.

- Let X be any sequence obtained from S that contains (a) all of the reals r_i in those ordered pairs p_i in S (which is to say, all of the reals in I) and (b) nothing else.

---

[9] To say it again, sequences are countable by definition, so we can't just extend a sequence to include an uncountable set of values (at least, not if the result is supposed to be a sequence still). On the other hand, we certainly can extend a sequence to include a *countable* set of values (and the result will still be a sequence), because "countable plus countable is countable." (This latter is a theorem. But the proof is straightforward, and I leave it as an exercise.)

[10] I remind you from Part I of this essay that a *one to one onto mapping* (also known as a *strict one to one correspondence* or a *bijection*) from set *s1* to set *s2* is a mapping, or function, such that each element of *s2* is the image under that mapping of exactly one element of *s1*. I remind you also that if M is such a mapping, then it has an inverse mapping M' from *s2* to *s1* that's also one to one and onto, and the phrase *one to one onto mapping between* is often used, and is used in this essay, to refer to such a mapping M and its inverse M' considered in combination.

- By Cantor's Interval Theorem, X can't exist. So S can't exist. So M can't exist. So the reals in I aren't countable. So the reals in their entirety aren't countable either, a fortiori. QED.

SOME QUESTIONS

There's one point regarding Cantor's Interval Theorem—a point in its favor, I hasten to add—that might or might not have occurred to you, but in any case I think is worth spelling out explicitly: namely, that the proof of that theorem differs from Cantor's diagonal proof (or "proof") inasmuch as it clearly has to do with numbers as such, not with their representations. But several other points occur in connection with that proof—points that, by contrast, don't exactly seem to be in its favor—that I think are therefore also worth raising, and that's the purpose of the present section.

Is Cantor's Proof Constructive?

Consider the following crucial extract from the text of Cantor's interval proof as I gave it earlier:

> Now let x_i and x_j be the first two distinct elements of X that are contained in I_0 ... Without loss of generality assume $x_i < x_j$ (so $a_0 < x_i < x_j < b_0$). Let I_1 be the interval $(a_1...b_1)$, where $a_1 = x_i$ and $b_1 = x_j$.

This text can be interpreted as an instruction to search through the sequence X, looking for the first two elements a_1 and b_1 that satisfy the condition $a_0 < a_1 < b_1 < b_0$. But given that X is infinite, that search might never terminate!—that is, we might never find those two elements in finite time. Worse, we might have to go on and perform a similar search an infinite number of times in order to come up with that infinite ordered collection of nested intervals $I_0, I_1, I_2, ...,$ etc.—and thus the proof overall involves, at least potentially, an infinite number of infinite searches. [11] And yet the Wikipedia entry "Cantor's First Set Theory Article" (see Appendix B to this part of the essay) states explicitly that Cantor's proof is constructive! How can this be?

> *Aside:* In case you're unfamiliar with the term *constructive* as I'm using it here, let me quote Wikipedia again:

>> In mathematics, a **constructive proof** is a method of proof that demonstrates the existence of a mathematical object by creating or providing a method for creating the object. This is in contrast to a **non-constructive proof** (also known as an **existence proof** ...), which [merely] proves the existence of a particular kind of object without providing an example.

---

[11] As a concrete illustration of at least part of this point, let me remind you of the following (a repeat of text from the discussion of the first of the two examples in the section before last): "I_3 is the last interval to be constructed ... because $b_3 = 1.99999$, and there's no x in the sequence X that (a) is strictly less than that value 1.99999 and (b) hasn't already been used as a boundary value for some previous interval." Determining that there's no such x clearly requires an infinite search of the infinite sequence X.

And it's usual to require also that the process identified for creating the object in question be "finitary" (i.e., not to involve anything infinite) in order for the proof to be considered constructive. However, it's only fair to mention too that not all mathematicians insist on this latter requirement. *End of aside.*

I'd like to pursue these issues a moment longer. Here again is the text from the section "Cantor's Interval Proof" that deals with what I there called Case 2ii:

> If $a_\infty = b_\infty$ (in which case the corresponding interval I_∞ is empty), then $y = a_\infty = b_\infty$ itself is an element of I_0 not appearing in X. For if it did appear in X, then we would have $y = x_n$ for some n. But as the lemma below implies, $x_n \notin (a_n...b_n)$ for all n; so $y (= x_n = a_\infty = b_\infty) \notin I_n$. But a_∞ and b_∞ are certainly elements of every interval that precedes the interval $I_\infty = (a_\infty...b_\infty)$ in the ordered collection of constructed intervals, and so they're both elements of I_n in particular (i.e., $y \in I_n$). So we have a contradiction (viz., $y \in I_n$ and $y \notin I_n$). It follows that $y = a_\infty = b_\infty$ is an element of I_0 that doesn't appear in X.

However, now consider the following argument:

- All of the a's in the sequence $a_0, a_1, a_2, ...$ (apart from a_0, possibly) are elements of X, by definition.

- By contrast, the limit $y = a_\infty$ of that sequence might be an element of X or it might not. Of course, it's certainly an element of every interval in the ordered collection $I_0, I_1, I_2, ...$, up to but not including I_∞, again by definition. (It's not an element of I_∞, but that's hardly surprising, since in the case we're considering—the one referred to in the proof as Case 2ii— I_∞ is empty.)

- But I see no reason why it necessarily has to be the case that y *isn't* an element of X. Cantor says, correctly, that if y is an element of X then there has to be some n such that $y = x_n$, and hence such that $y \notin I_n$. But if we have to allow n to be infinite in connection with a_n and I_n—i.e., if we have to be allowed to talk about a_∞ and I_∞—why can't we allow n to be infinite in connection with x_n as well (i.e., in connection with the requirement that $y = x_n$ for some n), and thus admit that what we're looking for might be some kind of "x_∞"?

- And if we do allow n to be infinite in connection with x_n, and if such a state of affairs happens to be the case, then Cantor's procedure will never find that x_n—i.e., that element x_∞ of I_0 that's not also an element of X. So how can the proof be described as "constructive"?

I now observe that the scenario just outlined—i.e., y (or x_∞) being an element of every interval in the ordered collection $I_0, I_1, I_2, ...$, up to but not including I_∞—is exactly the situation illustrated by the second of the examples I gave in the section before last.[12] Just to remind you, the given sequence X in that example was 1.9, 2.1, 1.99, 2.01, 1.999, 2.001, ..., and it deliberately omitted the element 2. However, the search portion of Cantor's procedure would need to be executed an infinite number of times, each such search involving an infinite number of elements, before it can be ascertained that the particular element 2 is indeed missing. So what exactly can we conclude, logically, from such a state of affairs?

I'll have more to say about such matters in the final subsection of this section ("Diagonalization Redux?").

Why Isn't X in Ascending Order?

Here again is the way I stated earlier what it was that Cantor wanted to prove with his interval argument:

> *Given:* A nonempty interval $I_0 = (a_0...b_0)$ and an infinite sequence $X = x_1, x_2, ..., x_n, ...$, where $a_0, b_0, x_1, x_2, ..., x_n, ...$ are reals and we assume without loss of generality that X contains no duplicates (i.e., the x's are all distinct from one another).

> *To prove:* There exists a real $y \in I_0$ such that $y \notin X$.

Note in particular the assumption that the x's are all distinct. Given this state of affairs, it might have occurred to you that there's surely a very easy way to prove Cantor's result, as follows. Since (to repeat) the x's are indeed all distinct, it must be that for all i, j where $i \neq j$, either $x_i < x_j$ or $x_j < x_i$. So why not just do the following?—

- Without loss of generality (because it can be sorted first if necessary), assume that X is in ascending order—i.e., assume that $x_1 < x_2 < x_3 < ...$.

- Choose any pair of consecutive x's in X—x_i and x_{i+1}, say—that are contained in the given interval I_0—i.e., such that $a_0 < x_i < x_{i+1} < b_0$. (Of course, if no such pair can be found, then the desired result is already proved.)

- Then $\frac{1}{2}(x_i+x_{i+1})$ is clearly an element of I_0 that's not an element of X. (In fact, of course, every element x of I_0 that satisfies $x_i < x < x_{i+1}$ is an element of I_0 that's not an element of X.)

What's wrong with the foregoing argument? Well, here's one thing: If X contains an infinite number of elements, what it unfortunately doesn't contain—at least, not necessarily—is a smallest one (i.e., a "least element"); for example, the sequence 0.1, 0.01, 0.001, 0.0001, ... doesn't. Thus, the sort process won't necessarily work, and so we can't assume that X is in ascending order after all.

---

[12] As for the first of those examples, see the previous footnote.

Not to mention the fact that the idea of trying to sort an infinite sequence might seem impossibly unrealistic anyway—though I don't know that it's any more unrealistic than the idea of trying to find the first two elements a_1 and b_1 in such a sequence such that $a_0 < a_1 < b_1 < b_0$, where a_0 and b_0 are given; frankly, the two notions strike me as equally suspect.

Are the Rationals Countable?

Here's something else that might have occurred to you. Assume Cantor's interval proof is valid. Now let all references to real numbers in that proof be replaced by references to rational numbers ("rationals") instead. Wouldn't the result be a proof that the rationals are uncountable? To spell it out, that result would be:

> Let I be a rational interval; then there's no rational sequence X that contains every rational in I.[13]

But in fact the rationals are easily seen to be countable, as Cantor himself proved (see Appendix C to Part III of this essay).

Fortunately, the resolution of this apparent contradiction is straightforward: Replacing all references to real numbers in Cantor's interval proof by references to rational numbers *doesn't* in fact yield a proof (or "proof") that the rationals are uncountable. Why not? Well, consider the following text (an abbreviated and lightly edited version of a critical portion of the proof as I gave it earlier):

> Suppose the total number of intervals so constructed is infinite. If it is, then the sequence a_0, a_1, a_2, ... has a limiting value, because the elements of that sequence are of finite magnitude and strictly monotonically increasing; likewise, the sequence b_0, b_1, b_2, ... also has a limiting value, because the elements of that sequence are of finite magnitude and strictly monotonically decreasing. Let those limiting values be a_∞ and b_∞, respectively ... If $a_\infty = b_\infty$, then $y = a_\infty = b_\infty$ is an element of I_0 not appearing in X.

But if we're talking about the rationals instead of the reals, then the claim in the final sentence of this extract is no longer valid. The reason is as follows: Even though a_0, a_1, a_2, ... and b_0, b_1, b_2, ... are now all rationals, the limiting values a_∞ and b_∞ might not be (rational, that is)—and if they're not, then they won't be contained in I_0 in the first place. (What's more, if they're not rational, the fact that they're also not contained in any sequence X of rationals (a) isn't exactly surprising and (b) more important, doesn't prove anything. In particular, it certainly doesn't prove that the rationals in I_0 are uncountable.)

Here's a simple example to illustrate the point that a_∞ and b_∞ might not be rational. Consider the well known Fibonacci sequence—

$$1 \quad 1 \quad 2 \quad 3 \quad 5 \quad 8 \quad 13 \quad \cdots$$

---

[13] You might notice that I haven't defined the terms *rational interval* and *rational sequence*, but of course this omission is easily remedied (the definitions are straightforward).

—in which each element from the third onward is the sum of the two immediately preceding elements. Now consider the sequence obtained by replacing each element after the first by the ratio of it and its immediate predecessor:

```
1    1/1    2/1    3/2    5/3    8/5    13/8    ...
```

The elements of this latter sequence are certainly all rational. As is well known, however, their limit is

```
( 1 + √5 ) / 2
```

(the Golden Ratio), which is irrational.

Well Ordering

Let me come back to that business of ordering. There's another point I want to make, or at least another question I want to ask, in that connection.

By definition, the elements of a set have no particular ordering (thus, e.g., {a,b,c} and {c,b,a} are both the same set).[14] However, we can and often do refer to the combination of

a. Some specific set, together with

b. Some specific ordering rule for the elements of that set (e.g., "order by ascending value"),

as "an ordered set"—and the reals in particular are an example of a *totally* ordered set, which just means that given any two distinct reals *a* and *b*, together with the usual arithmetical definition of "less than" ("<"), either *a* < *b* is true and *b* < *a* is false or the other way around. [15]

However, the reals don't form what's called a *well* ordered set. Here's a definition of this latter concept:

> **Definition (well ordered):** A set is well ordered if and only if it's totally ordered and every nonempty subset has a least element according to that ordering.

Thus, the set *R* of all reals, though it's totally ordered, isn't well ordered, because it has at least one nonempty subset—viz., the set *R* itself, which is certainly a

---

[14] I don't mean by this observation that it's not possible to impose an ordering on the elements of a set—it might be, and it might not (as in fact the very next sentence in the main text makes clear). Nor do I mean that the elements of, e.g., a sequence or an interval can't be considered as a set, if we ignore the ordering. All I mean is that the concept of ordering has no part to play in the concept, or definition, of a set as such.

[15] For the purpose of the present subsection I'm dropping my usual assumption that the reals we're talking about are all nonnegative; thus, by the term "reals" here I mean *all* reals, both negative and nonnegative.

subset of **R**, albeit not a proper one—that has no least element (there's no "smallest real").

On the other hand, there's also a theorem, the *Well Ordering Theorem*:

Theorem (well ordering): Every set can be well ordered.[16]

So if this theorem is legitimate, doesn't it contradict Cantor's claim that the reals are uncountable? To spell the point out:

■ Again let **R** be the set of all reals, but now ordered in accordance with the Well Ordering Theorem. Pair the first real in **R** with the natural number 1.

■ Remove that first real from **R** (and reorder what remains if necessary in accordance with the Well Ordering Theorem) to produce the ordered set **R'**, say. Pair the first real in **R'** with the natural number 2.

And so on. Since we're never going to run out of natural numbers, surely this procedure shows that every real can be paired with some unique natural number, and hence that there can't be "more" reals than natural numbers.

As a matter of fact I'm not sure we even need to bring well ordering into the picture anyway. All we need to be able to do is choose *some* real, out of an infinite set of reals[17]—not necessarily the "first," if "first" even has any meaning in this context—and remove it. The procedure outlined above becomes:

■ Let **R** be the set of all reals (not ordered in any particular way). Choose some real *r* in **R** and pair it with the natural number 1.

■ Remove *r* from **R** to produce the set **R'**, say; choose some real *r'* in **R'** and pair it with the natural number 2.

And so on.

What's more, suppose we interchange the roles of the reals and the natural numbers in the foregoing procedure. What results is something like this:

---

[16] Yes, I know, there does seem to be some contradiction here—apparently the set of all reals has no least element, and yet it can be well ordered, in which case it does have a least element after all (least according to that ordering, that is). Well, personally, I don't have the background knowledge to resolve this conflict satisfactorily. Let me just stress the point that even if there's something wrong with my argument in this particular subsection, all of the other arguments I'm offering (anti Cantor arguments, I mean) still stand.

[17] I choose the word "choose" deliberately, because I'm making a tacit appeal here to the *Axiom of Choice*, which can be stated as follows: Given a set S of nonempty, pairwise disjoint sets $s_1, s_2, ..., s_n$, there exists a set of n elements $x_1, x_2, ..., x_n$ such that each x_i is an element of s_i ($i = 1, 2, ..., n$). This axiom implies among other things that, given some set s, it must be possible to choose an arbitrary element x from that set (hence the name). *Note:* The Axiom of Choice is obviously and intuitively valid (and noncontroversial) so long as the sets $s_1, s_2, ..., s_n$, and S are all finite, but can be and has been questioned otherwise. On the other hand, it turns out that the Axiom of Choice is logically equivalent to the Well Ordering Theorem (in other words, each is valid if and only if the other is); thus, there might be some circular reasoning going on here. But if so, then I fall back on my position as articulated in the previous footnote, viz: Even if there's something wrong with my argument in this subsection, all of the other arguments I'm offering (anti Cantor arguments, I mean) still stand.

- Let N be the set of all natural numbers. Choose some natural number n in N (the smallest, perhaps) and pair it with some real r in R.

- Remove n from N to produce the set N', say; remove r from R to produce the set R', say; choose some natural number n' in N' (again the smallest, perhaps) and pair it with some r' in R'.

And so on. Since we're never going to run out of real numbers, surely this procedure shows that every natural number can be paired with some unique real, and hence there can't be "more" natural numbers than reals. So there are no more natural numbers than reals, and no more reals than natural numbers. So doesn't this argument show there must exist a one to one onto mapping between N and R—implying, by definition, that the reals are countable?

> *Aside:* Indeed, I believe the foregoing argument does show that such a mapping must exist. However, it doesn't explicitly *define* such a mapping; in other words, the argument is an existence proof merely, not a constructive one. Moreover, the argument relies, tacitly, on the Schröder-Bernstein Theorem, which states that if *s1* and *s2* are sets and there's a one to one into mapping (an injection) from *s1* to *s2* and a one to one into mapping (another injection) from *s2* to *s1*, then there's a one to one onto mapping (a bijection) between *s1* and *s2*. *End of aside*.

How Are Real Sequences Actually Defined?

Regardless of whether Cantor's Interval Theorem is in fact valid, it seems to me that the way it's usually stated isn't the most logically helpful, or appropriate. Let me explain. In English translation (see Appendix A to this part of the essay), that formulation begins thus:

> Given any definition of an infinite sequence of mutually distinct real numbers ...

Similarly, the Wikipedia entry "Cantor's First Set Theory Article" (see Appendix B to this part of the essay) says this:

> The theorem states: Given any sequence of real numbers $x_1, x_2, x_3, \ldots$

And here's the way I stated it myself earlier in the present essay:

> Let I be a real interval; then there's no real sequence X—no infinite real sequence X in particular—that contains every real in I.

What these various formulations all have in common is this: Each of them assumes that the concept of an arbitrary sequence of real numbers is well defined and doesn't involve any traps or hidden difficulties. But is such an assumption warranted? Well, consider the following:

- Let X be such a sequence.

- If X is finite, then it can obviously be defined by simple enumeration. For reasons noted earlier, however, the finite case isn't very interesting, and I'll ignore it for the remainder of this discussion.

- By contrast, if X is infinite, then the *only* way it can it be defined is by means of some rule, or in other words some algorithm or "generating function" f that gives the nth element of the sequence for all positive integers n.

> *Aside:* In this connection, I note again that Cantor's interval proof as given in Appendix A to this part of the essay begins "Given *any definition of* an infinite sequence" (italics added). The definition in question—i.e., the one referenced in that phrase "any definition of"—surely has to be understood as being in terms of some rule (in the sense in which I'm using that term *rule* here), precisely because the sequence in question is infinite and can't be defined by simple enumeration.
>
> Recall also the following remarks from one of the critics quoted in Part I of this essay:
>
>> But this is the "magic" of mathematics, if you like—that it's sufficient to have a formula or an algorithm for generating any given element, and thereby arbitrarily many elements, of a sequence, which in fact is how we *define* a countably infinite sequence.
>
> Well, I'm glad to say that on this point at least, if not on others, the critic and I are in agreement—that is, we agree that an infinite sequence, if it's to be defined at all, does indeed have to be defined in terms of "a formula or algorithm" (or rule, or generating function, or whatever else you might care to call it).[18]
>
> Consider also the following definition from the book *Infinite Series*, by James M. Hyslop (Oliver and Boyd, 1945):
>
>> Suppose that A_n is a function of the positive [integer] variable n which is defined for all values of n. Then the ordered set of numbers
>>
>> $$A_1, A_2, A_3, \ldots, A_n, \ldots$$
>>
>> obtained from A_n by giving n the values 1, 2, ... in turn is called an **infinite sequence** or, more simply, a **sequence**.
>
> I don't much care for either Hyslop's notation or his wording, but I do agree with him that, crucially, for an infinite sequence even to be defined in the first place, there needs to be some kind of generating function (which Hyslop denotes A_n).
>
> Finally, note the implications of all of the above for the hypothetical

---

[18] Admittedly, the critic says only that it's *sufficient* to have a generating function—but then, when he goes on to say "in fact [that's] how we *define* a countably infinite sequence," he makes it clear that such a function is necessary as well as sufficient.

list of all of the reals that Cantor's diagonal proof begins by assuming. To spell the point out, that list is certainly an infinite sequence; thus, Cantor's assumption is equivalent to assuming the existence of some generating function that produces all of the reals. *End of aside.*

■ So let f be a generating function for the infinite sequence X. Then if that function is such that it never evaluates to some specific real y (i.e., if there's no n such that $f(n) = y$), then that value y—which is therefore missing from the sequence X by definition—might be exactly the missing value referred to in Cantor's Interval Theorem (i.e., the value the theorem says is in the given interval I and not in the given sequence X).

■ And if that value y is indeed that missing value, then the fact that it's not contained in X (a) isn't exactly surprising and (b) more important, certainly doesn't prove that the reals in I are uncountable.

But Cantor's argument isn't that *some specific* sequence X will omit at least one real but that *all* such sequences X do. So the question becomes:

Does there exist a generating function f such that every real in the given interval I is produced by invoking f on some positive integer n?

If no such f exists, then a fortiori there can't be a sequence X that contains all of the reals in I. So it seems to me that Cantor's Interval Theorem would more appropriately be formulated in terms not of sequences but of generating functions—"There's no generating function f that generates all of the reals in I." And I'll show in Part III of this essay that this latter proposition is false, and hence that Cantor's Interval Theorem must be false as well.

Diagonalization Redux?

I believe the arguments in the foregoing subsections should be sufficient to warrant, at the very least, a serious degree of skepticism with respect to Cantor's Interval Theorem. In this, the final subsection of (the main body of) this part of the essay, I describe what I regard as a clinching argument in support of the position that the "theorem" in question doesn't prove what it claims to prove, and is thus not a theorem, as such, at all (hence the quote marks, of course).

I'll begin by repeating the following key insight from the section "Cantor's Interval Proof":

For all n > 0, if x_m is an element of X that appears in interval $I_n = (a_n...b_n)$, then x_m appears in X strictly to the right of b_n.

The following diagram (a minor variation on the one from the very end of the section "Cantor's Interval Proof") illustrates the point:

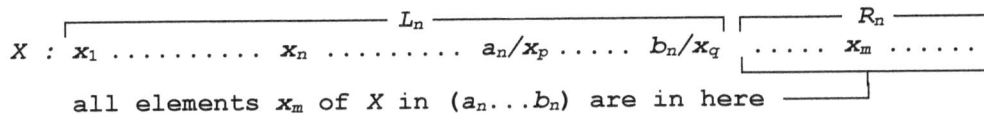

$$X : x_1 \ldots\ldots\ldots x_n \ldots\ldots\ldots a_n/x_p \ldots\ldots b_n/x_q \ldots\ldots x_m \ldots\ldots$$

with L_n spanning from x_1 to b_n/x_q, and R_n spanning over x_m

all elements x_m of X in $(a_n \ldots b_n)$ are in here

To elaborate: For all n, b_n is x_q for some q such that—as we saw in the section "Cantor's Interval Proof"—$q \geq 2n$. So we can imagine each such b_n as acting as a point of demarcation between two disjoint subsequences of X, L_n and R_n, where L_n consists of all elements of X to the left of b_n together with b_n itself, and R_n consists of all elements of X to the right of b_n. (Equivalently, L_n consists of all elements x_i of X where $i \leq q$, and R_n consists of all elements x_i of X where $i > q$.) So we can say that for all $y \in I_n$, if $y \in X$ then $y \in R_n$ (equivalently, for all $y \in I_n$, if $y \in X$ then y is x_m for some $m > 2n$).

Now, the sequence X consists of real numbers, and Cantor's Interval Theorem is supposed to apply to all possible X's. Moreover, since

a. The rationals are of course reals, and

b. As we know, the rationals are countable (again see Appendix C to Part III of this essay), and

c. "Countable plus countable is countable" (see footnote 9 to this part of the essay),

we're at liberty to assume the special case of an X that includes the rationals in their entirety (as well as, presumably, some collection—necessarily also countable—of irrationals).

Given such an X, then, it's easy to see that Cases 1 and 2i of Cantor's interval proof as I gave it earlier can't occur (which is why it's helpful to make the assumption, of course—it simplifies the overall argument). To elaborate:

■ For Case 1, the total number of constructed intervals was finite. But if X contains every rational, then given any nonempty interval $I_n = (a_n \ldots b_n)$, there will always exist rationals a_{n+1} and b_{n+1} in X such that $a_n < a_{n+1} < b_{n+1} < b_n$, because the rationals are "everywhere dense." (Again, see Appendix D to this part of the essay for an explanation of this term.) In other words, we can always construct the next nested interval I_{n+1}; so Case 1 can't now occur.

■ For Case 2i, the total number of constructed intervals was infinite, but a_∞ was strictly less than b_∞. But if $a_\infty < b_\infty$ there will always exist rationals a_{next} and b_{next} in X such that $a_\infty < a_{next} < b_{next} < b_\infty$—in which case those "$\infty$" suffixes on a_∞ and b_∞ are hardly very appropriate, by the way!—because (to say it again) the rationals are "everywhere dense." Thus, Case 2i also can't now occur.

So we're left with Case 2ii, in which $a_\infty = b_\infty$ and the corresponding interval I_∞ is empty. Let's think about that case. In the proof as I gave it earlier, the claim was that b_∞ (equivalently, a_∞) was itself an element of the original given interval I_0 that didn't appear in X. Now, this claim can't be valid if b_∞ is rational, because under our assumption *every* rational is an element of X, by definition. So what if b_∞ is irrational? Well, here again is the argument I gave previously (but here abbreviated and edited slightly):

> In this case $y = b_\infty$ is itself an element of I_0 not appearing in X. For if it did appear in X, then we would have $y = x_n$ for some n. But $x_n \notin (a_n...b_n)$ for all n; so $y \notin I_n$. But b_∞ is certainly an element of every interval that precedes the interval $I_\infty = (a_\infty...b_\infty)$ in the ordered collection of constructed intervals, and so it's an element of I_n in particular (i.e., $y \in I_n$). So we have a contradiction (viz., $y \in I_n$ and $y \notin I_n$). It follows that b_∞ is an element of I_0 that doesn't appear in X.

However, what this argument overlooks is the following. As we've seen, any specific b_n is equal to x_q for some specific q. Now, it's certainly true of that specific b_n that it's not equal to any of $x_1, x_2, ..., x_{q-1}$. But, crucially, all of those elements $x_1, x_2, ..., x_{q-1}$ lie in what I referred to above as L_n (as does x_q, which is to say b_n itself, also). In other words, $x_1, x_2, ..., x_{q-1}$ all appear by definition in the set of elements of X that have been considered, and rejected, in the process of determining b_n. Note carefully, however, that at this point *the elements of R_n haven't yet been considered at all.*

Let me say this again in different words, because it's important. We know that for all finite n ($n > 0$), the following are all true statements:

$b_n \in I_0$
$b_n \in I_1$
$b_n \in I_2$

...

$b_n \in I_{n-1}$
$b_n \notin I_n$

So what happens as n goes to infinity? Well, we can't assume the limiting value b_∞ appears in X, of course (see footnote 6 to this part of the essay); indeed, that's the very point at issue. Nevertheless, we can still regard that value b_∞ as a point of demarcation between two disjoint subsequences L_∞ and R_∞, consisting of those elements of X that have and haven't been considered, respectively. And it's still the case that b_∞ is an element of every interval in the sequence of intervals $I_0, I_1, I_2, ...,$ except for I_∞ itself ("the last one," if you see what I mean)—though the observation that $b_\infty \notin I_\infty$ is vacuous, of course, since I_∞ is empty. So far, then, the argument certainly doesn't show that $b_\infty \notin X$; in fact, it might even suggest the opposite—viz., that b_∞ does appear in X—since, to say it again, b_∞ certainly appears in every one of those constructed intervals except "the last one," which is empty anyway.

At this point, however, a Cantorian will argue that if b_∞ does appear in X, then it must be equal to x_q for some q, and so it won't appear in any interval I for which x_q is defined to be one of the boundary values. (Nor will it appear in any interval that's strongly included in that interval I, of course.) All of which is true enough: *but what hasn't been shown is that such an interval I is, and must be, one of the nested intervals produced by the construction process that Cantor's argument appeals to.* What's more, not only has the foregoing not been shown, but in fact it *can't* be shown—at least, not by that construction process—if that x_q happens to appear in R_∞, because the elements of R_∞ are never considered during that process.

So does b_∞ appear in R_∞? *Answer:* We don't know.[19] We don't know because, by the time we get to I_∞, it's still the case that the elements of R_∞ haven't been considered. To show that Cantor's result is valid, it would need to be shown that either (a) R_∞ is empty or (b) if it isn't, then there's some other reason why it can't contain b_∞. Since neither (a) nor (b) has been shown, what Cantor's Interval Theorem allegedly proves remains unproven.

Now, you might be tempted to object at this point that the recursive process of constructing those nested intervals is surely exhaustive; that is, by the time we "get to infinity" as it were, all of the elements of X will have been considered, and R_∞ will therefore indeed be empty. (Of course, for finite n, not only is R_n not empty, it's actually infinite. I mean, it's of infinite cardinality.) However, I think there might be a certain sleight of hand going on here (unintentional, I'm sure, but significant nonetheless). To be specific, I think those "infinite" suffixes on the symbols a_∞, b_∞, and so on, are misleading. To be even more specific, I think they suggest the following:

a. By the time interval I_n is considered, certainly elements $x_1, x_2, ..., x_n$ have all been considered and rejected.[20]

b. Therefore, by the time interval I_∞ is considered, elements $x_1, x_2, ...,$ up to and including a hypothetical "final" element x_∞, have all been considered and rejected.

c. Therefore R_∞ must be empty.

Now, point a. here is clearly valid. However, points b. and c. have not been shown (i.e., no logical reason has been given for believing them to be valid).

One last point: For any given n, let's agree to refer to the sequences L_n and R_n as the initial sublist and the remainder sublist, respectively. What happens as n goes to infinity? Well, the idea that a given value b_∞ doesn't appear in the initial sublist L_∞ but does appear, or might appear, in the remainder sublist R_∞ is very reminiscent of what happens with diagonalization as described in Part I of

_____

[19] Though if it turns out that Cantor's interval proof is just diagonalization in disguise—see the very last paragraph in this section—then in fact we would know (i.e., we'd know that b_∞ does appear in R_∞). *Note:* In case you're thinking the diagram near the beginning of this section surely shows that b_∞ *doesn't* appear in R_∞—after all, it does show that b_n doesn't appear in R_n—let me state for the record that the diagram in question is only an informal visual aid. It's not a formal proof. Note more particularly that what it does show is elements of X, and (to repeat) we can't assume up front that b_∞ is such an element. And note finally that what it also does show is that b_n is an element of L_n; so if we were allowed to conclude from this state of affairs that b_∞ is an element of L_∞, it would follow immediately that b_∞ is an element of X after all!

[20] Actually elements $x_1, x_2, ..., x_{2n}$ (and maybe more, but all of them appearing in X to the left of a_n and hence to the left of b_n as well, a fortiori) have been considered and rejected.

this essay. After all, diagonalization also involves the notion that a certain value doesn't appear in a certain preliminary portion (viz., the initial square sublist) of a certain sequence—actually a Cantor list (see footnote 11 to Part I of this essay)—but does appear later in that same sequence (viz., in the remainder sublist). So is it possible that Cantor's interval proof is basically just diagonalization in disguise? If it is, then all of the criticisms of diagonalization discussed in Part I of this essay (also in Part III, q.v.) will apply here also.

APPENDIX A: CANTOR'S ORIGINAL PROOF

Here for the record is Meyer's translation of Cantor's interval proof:[21]

Given any definition of an infinite sequence of mutually distinct real numbers,

$$(4)\, \omega_1, \omega_2, \ldots \omega_v, \ldots$$

then in any given interval $(\alpha \ldots \beta)$ there is a number η (and consequently infinitely many such numbers) where it can be shown that it does not occur in the series (4);[22] this shall now be proved.

We go to the end of the interval $(\alpha \ldots \beta)$, which has been chosen arbitrarily and in which $\alpha < \beta$; the first two numbers of our sequence (4) which lie in the interior of this interval (with the exception of the boundaries), can be designated by α', β', letting $\alpha' < \beta'$; similarly let us designate the first two numbers of our sequence which lie in the interior of $(\alpha' \ldots \beta')$ by α'', β'' and let $\alpha'' < \beta''$; and in the same way we determine the next interval $(\alpha''' \ldots \beta''')$ and so on. It follows that α', α'', … are by definition determinate numbers of our sequence (4), whose indices are continually increasing; the same applies for the sequence β', β'', … ; furthermore, the numbers α', α'', … are always increasing in size, while the numbers β', β'', … are always decreasing in size. Of the intervals $(\alpha' \ldots \beta')$, $(\alpha'' \ldots \beta'')$, $(\alpha''' \ldots \beta''')$, … each encloses all of those following.

There are two conceivable cases:

Either the number of intervals formed is finite; in which case, let the last such interval be $(\alpha^{(v)} \ldots \beta^{(v)})$. Since in its interior there can be at most one number of the sequence (4), a number η can be chosen from this interval which is not contained in (4), thereby proving the theorem for this case.

Or the number of intervals formed is infinite. Then the numbers α, α', α'', … because they are always increasing in size without growing infinitely large, have a determinate limiting value α^{∞}; the same holds for the numbers β, β', β'', …

---

[21] Copyright © 2018 *www.jamesrmeyer.com*. *Note:* I give Meyer's translation verbatim except for a couple of tiny editorial changes—I've replaced an "also" by "[if]" and a "so" by "[then]," and I've added a "[*sic*]."

[22] Don't let that "(4)" confuse you—it's just a label (the sequence so labeled just happens to be the fourth sequence or other formal construct explicitly called out in Cantor's paper). By the way, observe that Meyer's translation refers to that sequence as a "series," but it's clearly the former, not the latter.

because they are always decreasing in size, have a limiting value be [*sic*] β^{∞}; [if] $\alpha^{\infty} = \beta^{\infty}$ (such a case always occurs with the set (ω) of all real algebraic numbers), [then] one is readily satisfied, by looking back at the definition of the intervals, that the number $\eta = \alpha^{\infty} = \beta^{\infty}$ cannot be contained in our sequence; but if $\alpha^{\infty} < \beta^{\infty}$ then every number η in the interior of the interval $(\alpha^{\infty}...\beta^{\infty})$ or at its endpoints satisfies the requirement that it not be contained in the sequence (4).

Aside: With respect to that rather breathless final sentence (which constitutes almost the entirety of the paragraph beginning "*Or*"), the original text has a footnote reference attached to the assertion in line 7 to the effect that "the number $\eta = \alpha^{\infty} = \beta^{\infty}$ cannot be contained in our sequence." The footnote in question reads as follows:

> If the number η were contained in our sequence, then one would have $\eta = \omega_p$, where p is a specific index. But this is not possible, for ω_p does not lie in the interior of the interval $(\alpha^{(p)} ... \beta^{(p)})$, while by definition the number η does lie in the interior of the interval.

> Well, I don't know about you, but I don't think it's immediately obvious that what's being asserted in this footnote is in fact the case. That's why, in my own version of the proof of the theorem, I included a proof (a somewhat nontrivial proof, come to that) of that assertion as a lemma. In fact, whoever wrote the Wikipedia entry "Cantor's First Set Theory Article"—see Appendix B below—presumably felt the same way, because whoever it was did the same thing (included a corresponding lemma, I mean). *End of aside.*

APPENDIX B: THE WIKIPEDIA PROOF

The material of this appendix is based on, and consists in large part of excerpts from, the Wikipedia entry "Cantor's First Set Theory Article," which begins thus:

> **Cantor's first set theory article** contains Georg Cantor's first theorems of transfinite set theory, which studies infinite sets and their properties. One of these theorems is his "revolutionary discovery" that the set of all real numbers is uncountably, rather than countably, infinite. This theorem is proved using **Cantor's first uncountability proof**, which differs from the more familiar proof using his diagonal argument. The title of the article, "**On a Property of the Collection of All Real Algebraic Numbers**" ("Über eine Eigenschaft des Inbegriffes aller reellen algebraischen Zahlen"), refers to its first theorem: the set of real algebraic numbers is countable.

> [*text omitted*]

Cantor's second theorem works with a closed interval $[a, b]$, which is the set of real numbers $\geq a$ and $\leq b$.[23] The theorem states: Given any sequence of real numbers $x_1, x_2, x_3, \ldots$ and any interval $[a, b]$, there is a number in $[a, b]$ that is not contained in the given sequence. Hence, there are infinitely many such numbers.

"Cantor's second theorem" is, of course, the theorem I've been discussing at length in this part of the essay, and the Wikipedia article goes on to give the following as a proof of it:

To find a number in $[a, b]$ that is not contained in the given sequence, construct two sequences of real numbers as follows: Find the first two numbers of the given sequence that are in the open interval (a, b).[24] Denote the smaller of these two numbers by a_1 and the larger by b_1. Similarly, find the first two numbers of the given sequence that are in (a_1, b_1). Denote the smaller by a_2 and the larger by b_2. Continuing this procedure generates a sequence of intervals (a_1, b_1), (a_2, b_2), (a_3, b_3), ... such that each interval in the sequence contains all succeeding intervals—that is, it generates a sequence of nested intervals. This implies that the sequence $a_1, a_2, a_3, \ldots$ is increasing and the sequence $b_1, b_2, b_3, \ldots$ is decreasing.
Either the number of intervals generated is finite or infinite. If finite, let (a_L, b_L) be the last interval. If infinite, take the limits $a_\infty = \lim_{n \to \infty} a_n$ and $b_\infty = \lim_{n \to \infty} b_n$. Since $a_n < b_n$ for all n, either $a_\infty = b_\infty$ or $a_\infty < b_\infty$. Thus, there are three cases to consider:

Case 1: There is a last interval (a_L, b_L). Since at most one x_n can be in this interval, every y in this interval except x_n (if it exists) is not contained in the given sequence.

Case 2: $a_\infty = b_\infty$. Then a_∞ is not contained in the given sequence since for all n: a_∞ belongs to the interval (a_n, b_n) but x_n does not belong to (a_n, b_n). In symbols: $a_\infty \in (a_n, b_n)$ but $x_n \notin (a_n, b_n)$.

Case 3: $a_\infty < b_\infty$. Then every y in $[a_\infty, b_\infty]$ is not contained in the given sequence since for all n: y belongs to (a_n, b_n) but x_n does not.

The proof is complete since, in all cases, at least one real number in $[a, b]$ has been found that is not contained in the given sequence.

The Wikipedia entry also goes on to give the following as a proof of the lemma, used in Cases 2 and 3, to the effect that, for all n, $x_n \notin (a_n, b_n)$:

---

[23] As you can see, Wikipedia uses a notation for intervals that's slightly different from Cantor's (which is what I've used elsewhere in this essay). More to the point, note the Wikipedia claim that Cantor's proof is formulated in terms of closed (or "closed / closed") intervals of the form $[a, b]$. By contrast, the translation of that proof given in Appendix A to this part of the essay uses open (or "open / open") intervals of the form (a, b). (So does the body of this part of the essay, of course.) Whether this discrepancy has any significant consequences I leave as a matter for you to decide.

[24] The switch here from closed back to open intervals is a trifle disconcerting, to say the least. Whether it conceals some deeper problem I again leave as a matter for you to decide.

This lemma ... is implied by the stronger lemma: For all n, (a_n, b_n) excludes x_1, ..., x_{2n}. [25] This is proved by induction.

Basis step: Since the endpoints of (a_1, b_1) are x_1 and x_2 and an open interval excludes its endpoints, (a_1, b_1) excludes x_1, x_2.

> *Aside:* Wikipedia says here that "the endpoints of (a_1, b_1) are x_1 and x_2."
> No, they're not—not necessarily (see the section "Examples" earlier for a couple of counterexamples). Rather, the boundary values (or "endpoints," to use Wikipedia's term) of that interval are "the first two numbers of the given sequence that are in the open interval (a, b)," and there's no guarantee that "those first two numbers" are at any specific position whatsoever within the given sequence X.
>   That said, however, I observe that in a way the error doesn't matter. A corrected version of the erroneous statement would be "the endpoints of (a_1, b_1) are x_i and x_j for some $i \geq 1$ and some $j > i$." But if $i > 1$ instead of $i = 1$, then for different reasons—what reasons, exactly?—neither x_1 nor x_2 will be contained in (a_1, b_1), and hence it'll be "even more true," as it were, that x_1 in particular won't be an element of (a_1, b_1). *End of aside.*

Inductive step: Assume that (a_n, b_n) excludes x_1, ..., x_{2n}. Since (a_{n+1}, b_{n+1}) is a subset of (a_n, b_n)[26] and its endpoints are x_{2n+1} and x_{2n+2}, (a_{n+1}, b_{n+1}) excludes x_1, ..., x_{2n} and x_{2n+1}, x_{2n+2}. Hence, for all n, (a_n, b_n) excludes x_1, ..., x_{2n}. Therefore, for all n, $x_n \notin (a_n, b_n)$.

> *Aside:* This step is subject to a similar criticism. To be specific, Wikipedia says here that the endpoints of (a_{n+1}, b_{n+1}) are x_{2n+1} and x_{2n+2}, respectively, but again such is not necessarily the case. A corrected version of the statement would be that the endpoints of that interval are x_i and x_j. for some $i \geq 2n+1$ and some $j > i$. Again, however, the error doesn't really matter (right?). *End of aside.*

APPENDIX C: POTENTIAL vs. COMPLETED INFINITY

Ever since Aristotle (c. 384-322 BCE), mathematicians have drawn a distinction—a crucial distinction—between two kinds or "flavors," as it were, of infinity: viz., *potential* infinity vs. *completed* (or *actual*) infinity. This appendix explains and elaborates on such matters.

---

[25] The standard mathematical convention is to distinguish carefully between *containment* and *inclusion*—specifically, to say of a set that it *contains* its elements but *includes* its subsets. Thus, to say that some set *s1* "excludes" some set *s2* would, I presume, normally be taken to mean that *s1* doesn't include *s2* and hence that *s2* isn't a subset of *s1* (though "exclude" hardly seems the most apt verb for such a notion). In fact, however, Wikipedia is using *exclude* here to mean not *does not include* but *does not contain*—which I suppose is at least consistent (?), in a way, because Wikipedia also uses *contain* to mean *include*, when it says that "each interval in the sequence contains all succeeding intervals."

[26] Here Wikipedia is using the phrase "is a subset of"—very sloppily, I'd say—to mean "is strongly included in."

Potential Infinity

Potential infinity is a concept we appeal to all the time in mathematics, typically in the form of an ellipsis ("...") or words such as "etc." or "and so on"; it typically has to do with some process that starts out with some definite (and thus certainly finite) sequence of elements and goes on to perform some operation repeatedly—i.e., "an infinite number of times," potentially—on the elements of that sequence, thereby generating further elements and extending the sequence accordingly. For example, if we start with a sequence containing just the element 1/2 (one half) as "the current element" and the operation we perform repeatedly is "halve the current element and then extend the sequence by appending the result of that halving as the next current element," we obtain the sequence

```
1/2    1/4    1/8    1/16    1/32    . . .
```

Observe that the following properties hold with regard to this sequence (and indeed with regard to every sequence that's obtained by repeatedly invoking some "generating function"—see the body of this part of the essay—which is to say, *every possible* sequence):

a. Every individual element is finite. To spell the point out explicitly: No element is infinitely large (obviously not, in the example), and no element is infinitely small or "infinitesimal," either.

b. Every individual element is determined in a finite number of steps.

c. There's no final element, because the process of computing the next element (and extending the sequence accordingly) never ends. Instead, that rather loose—and in fact logically nonsensical—notion of "the final element" is replaced by the logically precise notion of a *limit* or *limiting value* (see below).

With the foregoing by way of informal introduction, let's take a closer and more formal look at such matters. I'll begin by considering functions. Here's a precise definition of what we mean when we say some function f goes to—or *tends* to, or *approaches*—a limit (k, say) as its sole parameter x goes to infinity:

> **Definition (limit of $f(x)$ as x goes to infinity):** Let f be a function with sole parameter x, where x denotes a real number. If and only if there exists a real number k such that, for all $\varepsilon > 0$, there exists $y > 0$ such that $x > y$ implies $|f(x)-k| < \varepsilon$, then k is the limiting value (or just limit for short) of $f(x)$ as x goes to infinity.[27] In symbols:

---

[27] Here and throughout this appendix I use notation of the form "| *exp* |" not (as elsewhere in this essay) to denote the cardinality of some set, but rather to denote the absolute value of the numeric expression *exp*. (The absolute value of some numeric value v is just v if v is nonnegative, the negation of v otherwise.)

$$\lim_{x \to \infty} f(x) = k$$

Aside: "Limit as x goes to infinity" is actually just a special case (though it's the case that's of primary interest to us here). The more general case is as follows:

Definition (limit of $f(x)$ as x goes to c): Let f be a function with sole parameter x, where x denotes a real number. If and only if there exists a real number k such that, for all $\varepsilon > 0$, there exists $\delta > 0$ such that $|x-c| < \delta$ implies $|f(x)-k| < \varepsilon$, then k is the limiting value (or just limit for short) of $f(x)$ as x goes to the value c. In symbols:

$$\lim_{x \to c} f(x) = k$$

End of aside.

By way of example, let the function f be $1+(1/x)$, and let the legal values of the sole parameter x be the positive reals (i.e., $x > 0$). Then it should be obvious that as x goes to infinity the limit of $f(x)$ is 1—

$$\lim_{x \to \infty} f(x) = 1$$

—because as x gets larger and larger, so $1/x$ gets closer and closer to 0, and thus $f(x)$ gets closer and closer to 1. *Note:* Translating this informal statement into the terminology of the foregoing formal definition is left as an exercise.

Now let f be a function for which the permitted values of its sole parameter x are just the positive integers. As explained in the section "Some Questions" in the body of this part of the essay, then, f can be regarded as the generating function for some sequence X. As a consequence, the notion of f having a limit as x goes to infinity can be extended in an obvious way to apply to that sequence X. However, we can if we like define the notion of the sequence X as having a limit without explicitly mentioning f at all:

Definition (limit of a sequence): Let X be the sequence $x_1, x_2, ..., x_n, ...$. If and only if there exists a real number k such that, for all $\varepsilon > 0$, there exists $y > 0$ such that $x > y$ implies $|x_n-k| < \varepsilon$, then k is the limiting value (or just limit for short) of X as n goes to infinity. In symbols:

$$\lim_{n \to \infty} X = k$$

By way of example, let X be the sequence—

```
1/2   1/4   1/8   1/16   1/32   . . .
```

—and let Y be the sequence $y_1, y_2, ..., y_n, ...$, where (for all n) y_n is equal to the sum $x_1+x_2+ ... +x_n$ of the first n elements of X. Note that x_n and y_n can each be regarded as a function of n. Then the limit of X is the limit of x_n as n goes to infinity, which is zero:

$$\lim_{n \to \infty} X = \lim_{n \to \infty} x_n = 0$$

And the limit of Y is the limit of y_n as n goes to infinity, which is one:

$$\lim_{n \to \infty} Y = \lim_{n \to \infty} y_n = 1$$

(Again, interpreting these examples in terms of the formal definition is left as an exercise.)

Let me now observe that—as in fact I hope you've already realized for yourself—the foregoing definitions and examples nowhere involve any notion of some parameter of some function, or some element of some sequence, actually being "equal to infinity" (a notion that makes no logical sense). Rather, what those definitions do, as indeed the examples illustrate, is this: They make it possible to reason about what happens if some operation is repeated "an infinite number of times" (i.e., indefinitely), without suggesting in any way that infinity is itself just another number. It's not. It's a concept.

Completed Infinity

In very strong contrast to everything I've said in the previous subsection, the notion of *actual* or *completed* infinity involves the acceptance of infinite constructs as self-contained objects that actually exist and can be manipulated and reasoned about in their entirety. To elaborate briefly:

- Here are some examples of finite objects: (a) the set of all primes less than 1,000; (b) the circle with some specific center and radius; (c) the number of different ways of choosing three elements out of a set of cardinality ten.

- Here by contrast are some examples of infinite objects: (a) the set of all primes; (b) the set of all circles; (c) the number of different ways of choosing three elements out of an infinite set.

Aristotle rejected the notion of actual infinity because (to quote Wikipedia) "if it were possible, then something would have attained infinite magnitude, and would be bigger than the heavens." By contrast, he accepted the notion of potential infinity: "Mathematics relating to infinity is not deprived of its applicability by [rejecting the notion of actual infinity], because mathematicians do not need the infinite for their theorems, just a finite, arbitrarily large magnitude."

Cantor's Proofs

I now claim that (a) Cantor's diagonal proof as discussed in Part I of this essay, and (b) his interval proof as discussed in the present part, both rely, at least implicitly, on the notion of completed infinity—a notion, to reiterate, that Aristotle specifically rejected. To elaborate:

- *The diagonal proof:* As we saw in Part I of this essay, Cantor applies his diagonalization scheme to a Cantor list[28] to produce, for all n, a string that doesn't appear in the initial $n \times n$ square sublist but does appear in the remainder sublist. And he goes on to ask us to believe that when n becomes infinite—i.e., when we "complete the infinity"—then the foregoing state of affairs no longer holds. (Moreover, the strings in a Cantor list must themselves be understood as "completed infinities" also, in order for that infinite diagonalization process to apply in the first place. Insofar as it does apply, I suppose I should add.)

- *The interval proof:* Here Cantor defines a procedure according to which, for all $q > 2n$, (a) the qth element x_q of a given sequence X fails to appear in the last interval I_n in a nested collection of open intervals $I_0, I_1, I_2, ..., I_n$, but (b) for any given $m > q$, the mth element x_m of X might or might not appear in that same last interval. And he goes on to ask us to believe that when n becomes infinite—i.e., when we "complete the infinity"—then the foregoing state of affairs shows that X fails to contain all of the elements in those nested intervals. (This, despite the fact that "the interval at infinity" I_∞ actually contains no elements at all, and so no x_m can possibly appear in it, even if it appears—which it might—in every preceding interval in the collection.)

What's more, it seems to me that to assume, as Cantor's proofs apparently do, that "completed infinities" actually exist is rather close to assuming what those proofs are attempting to prove: viz., that transfinite numbers actually exist in turn.

A Remark on the Terminology

I have to say I don't find the terminology of potential vs. completed (or actual) infinity very helpful, or intuitive. Regarding "potential" infinity, I think a better way to characterize the situation is simply to say there's *no bound*, or limit, on (e.g.) the number of times a particular operation is to be performed. Here again is the example I used earlier to illustrate this concept:

a. Start with a sequence containing just 1/2 as the current element.

b. Perform the following operation repeatedly: "Halve the current element and then extend the sequence by appending the result of that halving as the next current element."

c. We obtain the following sequence:

$$1/2 \quad 1/4 \quad 1/8 \quad 1/16 \quad 1/32 \quad \cdots$$

---

[28] I remind you—see footnote 11 in Part I of this essay—that I use the term *Cantor list* to mean one of those hypothetical complete lists of reals that Cantor's diagonal argument begins by assuming. Of course, that's one "completed infinity" right there.

In this example, the operation of halving the current element and extending the sequence accordingly is understood as being performed an unbounded number of times. Likewise, the final sequence is understood to be of unbounded length. So *unboundedness* is the crucial aspect of what's going on here, and to my mind it would be clearer to state as much explicitly, instead of referring to "potential infinity"—which, when you come to think of it is a rather strange term anyway. I mean, surely something either is infinite or it isn't, and so to say it's "potentially" infinite seems to my way of thinking somewhat mealy mouthed, not to say confusing.

As for "actual" or "completed" infinity: Here the problem, to my mind, is simply that those qualifiers tend to suggest that infinity is just another number. As I said a couple of pages back, though, it's not a number, it's a concept. For that reason, I'd prefer to call it just infinity, unqualified.

APPENDIX D: ON THE NOTION OF "EVERYWHERE DENSE"

As noted in the section "Cantor's Interval Proof," the interval I and sequence X referred to in Cantor's Interval Theorem are both infinite, in the sense that they both contain an infinite number of reals—I by definition (because it's nonempty), and X because it's so stipulated. Here, however, I want to focus on the interval I in particular. The fact is, that interval isn't just infinite in the foregoing sense, it's *everywhere dense* (a concept that was in fact first defined by Cantor himself). Here's a definition:

> **Definition (everywhere dense):** Let S be a totally ordered set; i.e., let S be such that whenever p and q are distinct elements of S, either $p < q$ or $q < p$. Then S is everywhere dense—sometimes just dense for short—if and only if, whenever elements p and q of S are such that $p < q$, there's another element r of S such that $p < r < q$.

Points arising from this definition:

- If S is empty or contains just one element, then it's everywhere dense, but trivially so.

- If S is everywhere dense, then (to repeat) whenever elements p and q of S are such that $p < q$, there's another element r of S such that $p < r < q$. Of course, it follows from this state of affairs that there are also elements r' and r'' of S such that $p < r' < r$ and $r < r'' < q$; in other words, between any such p and q, there are in fact an infinite number of elements r of S. Thus, if S is everywhere dense and contains at least two elements, then it must in fact be infinite.

  Note that not all infinite sets are everywhere dense, even if "<" is defined for all pairs of elements of the set in question. A counterexample is N, the set of all natural numbers.

If S is everywhere dense, then we can extend the definition in an obvious way to say that any interval defined over elements of S is everywhere dense also. In particular, if S is R, the set of all reals, and the defined ordering is by ascending value, then every real interval—the given interval I referred to in Cantor's Interval Theorem in particular—is everywhere dense.

APPENDIX E: DASGUPTA'S PROOF

In his book *Set Theory with an Introduction to Real Point Sets*,[29] the author, Abhijit Dasgupta, offers another interval-based proof that the reals are uncountable. (More specifically, he shows that the reals in the closed interval [0...1] are uncountable, from which the more general result follows.) In outline, Dasgupta's proof runs as follows:

1. Let I_0 be the closed real interval [0...1], and let f be a function that maps positive integers to reals in I_0; also, for all $n > 0$ let the result of evaluating $f(n)$ be x_n. (Note that those results $x_1, x_2, x_3, ...$ constitute a sequence X, and further that f is the generating function for X.)

2. Split I_0 into three equal subintervals $L_0 = [0...\frac{1}{3}]$, $M_0 = [\frac{1}{3}...\frac{2}{3}]$, and $R_0 = [\frac{2}{3}...1]$; then discard M_0. (The only reason for bringing M_0 into the picture at all is to ensure that L_0 and R_0 are disjoint.) Clearly x_1 is an element of at most one of L_0 and R_0 (possibly of neither). Let I_1 be L_0 if $x_1 \notin L_0$ and R_0 otherwise.

3. Split I_1 into three equal subintervals L_1, M_1, and R_1, and discard M_1. Clearly x_2 is an element of at most one of L_1 and R_1. Let I_2 be L_1 if $x_2 \notin L_1$ and R_1 otherwise.

And so on. Continuing in this fashion, we obtain a collection of nested closed intervals $I_1, I_2, ..., I_n, ...$ (all of them subintervals of the original interval I_0) such that (a) $I_1 \supset I_2 \supset ... \supset I_n \supset ...$[30] and (b) for all n, $x_n \notin I_n$. Now we appeal to the Nested Interval Theorem, according to which there exists exactly one element y such that $y \in I_n$ for all n.[31] It follows that:

a. $y \neq x_n$ for any n.

---

[29] Springer Science+Business Media, 2014.

[30] The expression $I_i \supset I_j$ means I_j is properly included in I_i.

[31] You might be surprised by this claim, given that in the body of this part of the essay I argued that the final interval in a similar sequence of nested intervals had to be empty. But the intervals under discussion in that previous argument were open, whereas the intervals under discussion here are closed. There's no such thing as an empty closed interval. (As a matter of fact, the Nested Interval Theorem clearly implies that the final interval in the sequence $I_1, I_2, I_3, ...$ will contain just that element y and nothing else; but why, exactly?.)

b. Hence y is an element of the original interval I_0 that's not equal to $f(n)$ for any n.

c. Hence there are "more" reals in I_0 than there are positive integers.

d. Hence the reals in I_0, and more generally the reals in their entirety, are uncountable. QED.

As you can see, Dasgupta's proof is formulated in terms of intervals, not Cantor lists. Thus, you might be thinking the proof in question is just Cantor's interval proof in disguise, as it were. But it's not; rather, it's Cantor's *diagonal* proof in disguise. To demonstrate the truth of this claim, I'll go through Dasgupta's proof again—but this time I'll present it a little more carefully (and in slightly revised form), and I'll incorporate certain additional comments and steps of my own as we go.

Given: The closed / open real interval $I_0 = [0...1)$ and a function f that maps positive integers to reals in I_0 (in other words, $f(n) \in I_0$ for all integers $n > 0$).
  Note: As I pointed out earlier, evaluating $f(n)$ for $n = 1, 2, ...$ yields an infinite sequence $X = x_1, x_2, ...,$ such that for all n ($n > 0$) x_n is a real in I_0. So we can, and from this point forward I will, take that sequence X as our given instead of the function f. Also, the reason for switching from closed / closed intervals to closed / open ones will become clear in a few moments.

To prove: There exists $y \in I_0$ such that $y \notin X$; equivalently, there exists $y \in I_0$ such that for all $n > 0$, $f(n) \neq y$.

Construction: An ordered collection of intervals (all of them closed below and open above) $I_1 = [a_1...b_1)$, $I_2 = [a_2...b_2)$, ..., such that, for all $n > 0$, (a) I_n is properly included in I_{n-1} (i.e., $I_0 \supset I_1 \supset I_2 \supset ... \supset I_n \supset ...$) and (b) $x_n \notin I_n$. For further details, please read on.

Proof: Before getting into the specifics of the proof as such, it might help to point out the following. Let x be an arbitrary element, expressed in binary, of the sequence X. Then:

- If the first bit of x is:

    0 : then $x \in [0.0...0.5)$ – i.e., the left half of I_0
    1 : then $x \in [0.5...1.0)$ – i.e., the right half of I_0

  (interval boundary values shown in decimal for clarity).

- If the first two bits of x are:

> 00 : then $x \in [0.00..\,0.25)$ – i.e., the first quarter from the left of I_0
> 01 : then $x \in [0.25..\,0.50)$ – i.e., the second quarter from the left of I_0
> 10 : then $x \in [0.50..\,0.75)$ – i.e., the third quarter from the left of I_0
> 11 : then $x \in [0.75..\,1.00)$ – i.e., the last quarter from the left of I_0

And so on. So consider now the following procedure:

1. Split I_0 into subintervals $L_0 = [0.0...0.5)$ and $R_0 = [0.5...1)$,[32] and consider the first element x_1 of X. Clearly x_1 is an element of exactly one of L_0 and R_0. Let $y = y_1$, where $y_1 = 1$ if $x_1 \in L_0$ and 0 otherwise. Observe that y_1 is "the ones complement" of the first bit of x_1 (thus certainly $y \neq x_1$).

2. Let $I_1 = [a_1...b_1)$ be whichever of L_0 and R_0 does *not* contain x_1; i.e., if $y_1 = 0$ let I_1 be L_0, otherwise let I_1 be R_0. Split I_1 into subintervals $L_1 = [a_1...\frac{1}{2}(b_1-a_1))$ and $R_1 = [\frac{1}{2}(b_1-a_1)...b_1)$, and consider the second element x_2 of X. Clearly x_2 is an element of at most one of L_1 and R_1 (possibly of neither). Let $y_2 = 1$ if $x_2 \in L_1$ and 0 otherwise. Extend y by appending y_2 at the right (i.e., replace y by $y\,||\,y_2$, where "$||$" denotes string concatenation). Observe that y_2 is the ones complement of the second bit of x_2 (thus certainly $y \neq x_2$).

3. Let $I_2 = [a_2...b_2)$ be L_1 if $y_2 = 0$, or R_1 otherwise. (In other words, let I_2 be L_1 if $x_2 \notin L_1$ and R_1 otherwise. Observe that either way, we have $x_2 \notin I_2$.) Split I_2 into subintervals $L_2 = [a_2...\frac{1}{2}(b_2-a_2))$ and $R_2 = [\frac{1}{2}(b_2-a_2)...b_2)$, and consider the third element x_3 of X. Clearly x_3 is an element of at most one of L_2 and R_2 (possibly of neither). Let $y_3 = 1$ if $x_3 \in L_2$ and 0 otherwise. Replace y by $y\,||\,y_3$. Observe that y_3 is the ones complement of the third bit of x_3 (thus certainly $y \neq x_3$).

And so on. Continuing in this fashion we obtain a bit string y, which (for reasons that should soon become apparent, if they aren't so already) I'll refer to as the *antidiagonal* corresponding to the sequence X.

Now, at this point a Cantorian will presumably argue as follows:

- For all $n > 0$, x_n differs from y in the nth bit position; hence $y \notin X$.

- But certainly y represents a real in the interval $I_0 = [0...1)$. Thus, X doesn't contain all of the reals in I_0, and so it can't be the case after all that the given function f generates all of the reals in that interval.

- But there's nothing special about the interval I_0 or the function f, or equivalently the sequence X. In other words, given *any* nonempty real

---

[32] Now we see the reason for the switch to closed / open intervals—there's no need to split into three subintervals as Dasgupta does, because using closed / open instead of closed / closed guarantees that the left and right subintervals will be disjoint.

interval I and *any* sequence X of reals, there always exists an element y of I that doesn't appear in X; hence the reals in I (and, more generally, the reals in their entirety) are uncountable. QED.

So what if anything is wrong with the foregoing? The crucial point, it seems to me, lies in the conclusion that (to say it again) for all $n > 0$, x_n differs from the antidiagonal y in the nth bit position. Well, we've seen this kind of conclusion before. What's more, it's valid, as far as it goes. However, if we think of the sequence X as forming a Cantor list (which it clearly does), then *by definition* that conclusion applies only to a proper sublist of that list—namely, the portion I referred to in Part I of this essay as the initial square sublist. Let me spell the point out for the record:

- Dasgupta shows that for all $n > 0$ and for all m ($0 < m \leq n$), $y \neq x_m$; in other words, y doesn't appear in the first n elements $x_1, x_2, ..., x_n$ of X.

- But it doesn't follow that y doesn't appear in X at all! Specifically, it doesn't follow that there exists no m ($m > n$) such that $y = x_m$.

- To put the point more precisely: Let n be an arbitrary positive integer. Then what Dasgupta's argument shows is that no element in the sequence $x_1, x_2, ..., x_n$ has y as its first n bits. But that sequence $x_1, x_2, ..., x_n$ is only the initial subsequence of the complete sequence X! There could be as many as $2^n - n$ additional elements of X not contained in that subsequence, and Dasgupta's result has nothing whatsoever to say as to whether any of those additional elements has y as its first n bits. In fact, as n increases, Dasgupta's argument deals with less and less of the complete sequence X (because $2^n - n$ grows much faster than n does), and thus it becomes more and more likely, as it were, that some element of X does indeed have y as its first n bits.

To sum up, Dasgupta's proof is basically just Cantor's original diagonal proof in disguise, and it suffers from the same flaws as that proof does.[33]

Postscript

In case you're still not convinced, let me say a little more regarding the differences as I see them between Dasgupta's proof and Cantor's interval proof (his interval proof, let me stress, not his diagonal proof). It seems to me the biggest difference is this:

---

[33] One flaw you might have noticed for yourself: Like Cantor's diagonal proof (but *not* like his interval proof), Dasgupta's proof deals not with numbers as such but rather with their binary representations. As a consequence, the bit strings 100000... and 010000... (for example) both represent "one half," but one of them falls in the left half, and the other in the right half, of interval I_0. (To be fair, though, this flaw doesn't arise in connection with Dasgupta's original proof, where intervals are split into thirds, but only in connection with my revised version of that proof where they're split into halves.)

a. At each stage in his recursive procedure Cantor defines the next interval by *searching the sequence* X, looking for the first two elements that are contained in the interval defined during the previous iteration (and taking those two elements as the boundary values for that next interval). So the boundary values of Cantor's intervals—with the possible exception of the final interval I_∞—are always explicitly elements of X.

b. By contrast, at each stage in his recursive procedure Dasgupta defines the next interval by *splitting the previous one*—that previous interval being one that explicitly doesn't contain the previous element of the sequence X. There's thus no question of the boundary values of that next interval being required to be elements of X.

Here are some further points of difference:

■ As noted above, Dasgupta's proof includes an appeal to the Nested Interval Theorem, which Cantor's proof doesn't. What's more, that theorem depends on a certain *Axiom of Completeness*, which seems to be almost, albeit not quite, equivalent to the theorem in question! (At best the theorem is a fairly trivial corollary of the axiom.) To me this all looks rather dangerously close to simply assuming what the theorem supposedly proves.

Aside: For the record, the difference between the Axiom of Completeness and the Nested Interval Theorem is as follows: Given a collection of nested closed intervals, the axiom says there's at least one element that belongs to all of them; the theorem says there's exactly one that does. *End of aside*.

■ Cantor's proof requires a notion of what I've called *strong* inclusion (of intervals). Dasgupta's proof relies only on the weaker notion of *proper* inclusion.

■ Cantor's proof relies on the somewhat suspect assumption that we can search an infinite set looking for elements that satisfy some condition. The counterpart to this assumption in Dasgupta's proof is merely the much more reasonable assumption that we can tell whether some element x is an element of some specific interval (which requires only a couple of simple numeric comparisons, viz., between x and the boundary values of the interval in question).

■ Cantor's proof has to deal with the possibility that the set of constructed intervals might have finite cardinality (i.e., the recursion doesn't always have to "go to infinity"). Dasgupta's proof always goes to infinity.

■ Cantor's proof relies on the sequences of a's and b's having limiting values. Dasgupta's proof involves nothing analogous.

- Cantor's proof involves a slightly nontrivial lemma to show that the nth element of X isn't contained in the nth interval. Dasgupta's proof avoids the need for that lemma by construction.

Part III:

Case Closed

It is certainly not the least charm of a theory that it is refutable.
—Friedrich Nietzsche:
Beyond Good and Evil (1885-1886)

Cantor gave two separate proofs, or would-be proofs, of his claim that the reals are uncountable, one based on Cantor lists and diagonalization, the other on nested intervals. In Part I of this essay, I presented arguments that cast doubt on the first of those proofs (and indeed, and in some ways more important, on the whole idea of using diagonalization as a basis for any kind of proof at all); in Part II, I presented arguments that cast doubt on the second (and suggested, moreover, that in any case Cantor's interval argument might just be diagonalization in disguise). But after I'd completed what I thought at the time were the final drafts of those two parts, I came across a paper that as far as I'm concerned resolves the issue once and for all. The paper in question was *Contra Cantor: How to Count the "Uncountably Infinite,"* by Erdinç Sayan.[1] In that paper, Sayan presents:

a. Yet another argument to show that Cantor's diagonalization scheme is fundamentally flawed;

b. An argument to show in particular that ("contra Cantor") the reals are in fact countable after all; and

c. An argument to show that the specific theorem that's usually meant when people use the unqualified name "Cantor's Theorem"—which is to say, Cantor's theorem to the effect that the cardinality of the power set of a given set is strictly greater than that of the set in question—is wrong as well, if the given set happens to be infinite. (It's valid if the set in question is finite.)

In what follows, I'll consider each of these contributions in turn. Before I do, however, let me explicitly acknowledge the huge debt that everything in this part of my essay owes to Sayan's beautiful paper.

---

[1] *https://www.academia.edu/37229455/contra_cantor_how_to_count_the_uncountably_infinite_e. Note:* I must also give credit here to David McGoveran, who independently came up with arguments that turned out on closer examination to be equivalent to certain of those in Sayan's paper. The arguments in question were presented in McGoveran's paper "Diagonalization" (see footnote 2 in Part I of this essay). Indeed, not only was that paper—as noted earlier—the principal reference source for that first part of this essay, but in fact it predated the version of Sayan's paper that I'm considering here by at least three years.

DIAGONALIZATION REVISITED

Cantor's diagonal proof begins by considering a Cantor list, or in other words a hypothetical complete list of the reals in a given interval, typically the closed / closed interval [0...1]—and for definiteness I'll assume this specific case throughout what follows, barring explicit statements to the contrary. (If you need a reminder as to why it's acceptable to limit the investigation in such a manner, please see Part I of this essay.) Here are the first few entries in such a list (this example is taken from Sayan's paper):

```
 1.   1   0   1   1   1   0   1   1   0   0   0  ...
 2.   1   1   0   1   0   1   1   1   0   0   0  ...
 3.   1   1   1   0   0   1   0   0   1   1   0  ...
 4.   0   1   1   0   1   1   1   0   0   0   1  ...
 5.   1   0   0   0   0   1   0   0   0   1   1  ...
 6.   1   1   0   0   0   1   1   0   1   0   0  ...
 7.   1   0   0   1   0   0   0   1   1   1   0  ...
 8.   0   0   1   0   1   1   1   0   0   0   1  ...
 9.   0   1   1   1   1   0   1   1   1   0   0  ...
10.   0   0   1   1   1   0   0   1   0   1   1  ...
11.   0   1   0   1   0   0   0   1   0   1   0  ...
      etc.
```

Every bit string in this list is to be understood as denoting the fractional part, expressed in binary, of a real number x whose integer part is 0—which is to say, a real number in the range $0 \leq x \leq 1$, or equivalently a real number in the interval [0...1]. *Note:* For brevity, from this point forward I'll use expressions of the form "the real number x," where x is a certain bit string, as shorthand to mean that real number ("real" for short) whose integer part is 0 and whose fractional part is represented in binary by that bit string x—more specifically, whose fractional part is represented that way in some Cantor list.

Next I define what it means for two such reals to be *inverses*:

Definition (inverse reals): Let x be a real in the interval [0...1], and let x' be the unique real that has a 0 in just those positions where x has a 1 and vice versa. Then, and only then, x and x' are inverses of one another.

Observe that x' is also an element of that same interval [0...1], and hence that x and x' must both appear in any given Cantor list (because such a list is supposed to contain *all* of the reals in that interval). In the example above, entries 4 and 7 are inverses of each other, and so are entries 6 and 10.

Aside: *Inverse* is Sayan's term for this concept. *Ones complement*—sometimes just *complement*, unqualified—is a more conventional but clunkier term (and indeed I used that term in Appendix E to Part II of this essay). Here I'll stay with Sayan's term. *End of aside.*

Now let's apply Cantor's diagonalization scheme, or what in Part I of this essay I called the choice algorithm, to the example above. First, here's the example repeated, but now with the diagonal starting at top left highlighted:

```
 1.   1  0  1  1  1  0  1  1  0  0  0 ...
 2.   1  1  0  1  0  1  1  1  0  0  0 ...
 3.   1  1  1  0  0  1  0  0  1  1  0 ...
 4.   0  1  1  0  1  1  1  0  0  0  1 ...
 5.   1  0  0  0  0  1  0  0  0  1  1 ...
 6.   1  1  0  0  0  1  1  0  1  0  0 ...
 7.   1  0  0  1  0  0  0  1  1  1  0 ...
 8.   0  0  1  0  1  1  1  0  0  0  1 ...
 9.   0  1  1  1  1  0  1  1  1  0  0 ...
10.   0  0  1  1  1  0  0  1  0  1  1 ...
11.   0  1  0  1  0  0  0  1  0  1  0 ...
      etc.
```

As you can see, the diagonal is the string

```
1  1  1  0  0  1  0  0  1  1  0 ...
```

Call this string d. Clearly, d represents a real in the interval [0...1], so it must itself appear in the list, and so it does (it's entry number 3). Hence, its inverse (call it d')—

```
0  0  0  1  1  0  1  1  0  0  1 ...
```

—(in other words, the corresponding antidiagonal) must appear in the list as well. *But it can't.* Why not? Well:

■ Suppose it appears as the nth entry.

■ By definition, then, the diagonal d will intersect that nth entry at the nth bit position. (Assuming it intersects it all, that is! See Part I of this essay.)

■ So if the nth bit of that nth entry (i.e., the nth bit of the entry for d') is 0, the nth bit of the diagonal d must also be 0; so the nth bit of the inverse d' of d must be 1, and so the nth bit of the d' entry must be 1. In brief: If the nth bit of d' is 0, then it must be 1; and by a precisely analogous argument, of course, if it's 1, then it must be 0. Contradiction!

■ So d' can't appear as the nth entry for any n; in other words, it can't appear in the list at all.

Now, you might be thinking that all the foregoing argument does is show that Cantor was right after all, since we've simply followed Cantor's procedure and thereby constructed a number d' that doesn't appear in the given list. But Sayan sees it differently. Here again is that contradiction spelled out above, but now in slightly abbreviated form:

■ If the nth bit of d' is 0, it must be 1, and if it's 1 it must be 0. And since d is the inverse of d', the same goes for d as well.

Given this state of affairs, Sayan concludes that d' is a "non number" (because it involves a "paradoxical" bit), and therefore, to repeat, it can't appear in the list—and the same goes for d, of course. Here's an excerpt from Sayan's paper (though I've edited it slightly to make it fit more neatly into the present context):

> Cantor's faulty reasoning was to think that he could assume initially that the list was complete, apparently being oblivious to the complication stemming from such an assumption's requirement that d and its inverse d' must both be included in the list. The complication is, to repeat, the clash of the d' entry and the diagonal d at one digit ... which makes that digit of the diagonal paradoxical. Taking the inverse of a diagonal that embodies paradox, of course, doesn't make sense. Cantor's construction of d' would not have looked so straightforward and unproblematic if he had taken cognizance of this complication brought about by the completeness assumption. My and Cantor's reasons for why d' cannot be contained in the list are not the same. The difference between my position and Cantor's may sound subtle but is nevertheless real ... I claim that, because of the initial assumption of completeness, ... the diagonal of the list becomes unsuitable for Cantorian inversion, and d and d' become non numbers.
>
> *Aside:* One of my reviewers objected to the foregoing argument (the argument, that is, to the effect that d and d' aren't genuine numbers), saying it was "specious" and in effect nothing more than another way of saying that diagonalization applies only to the initial square sublist. But I don't quite see it that way; rather, I think Sayan is pointing out a different problem. To be specific, he's pointing out that (a) Cantor just assumed we're allowed to take the diagonal in the first place, without realizing that (b) that assumption necessarily entails the notion of the diagonal intersecting the entry for the antidiagonal at some specific bit position, with the contradictory consequence already explained. *End of aside.*

In other words, what Sayan's argument does is this: It identifies a new and fatal flaw in the process of diagonalization. Now, diagonalization in turn rests on the assumption that a Cantor list can be used to represent the reals in their entirety (or the reals within some interval in their entirety, at any rate); so that assumption is false, and no such list exists. As far as that conclusion is concerned, therefore (viz., that no such list exists), Cantor was right! But it doesn't follow from that conclusion that we can't enumerate, or count, the reals in some other way—and in fact we can, as the section immediately following demonstrates.[2]

UNCOUNTABILITY OF THE REALS REVISITED

Consider the following tree structure:

---

[2] Where Cantor went wrong—and where, it seems, so many others since Cantor have gone wrong as well—was to assume that countable implies listable (at least, "Cantor listable," meaning listable in the Cantor list sense). It's true that listable implies countable, but the converse is false. Cf. footnote 11 in Part I of this essay.

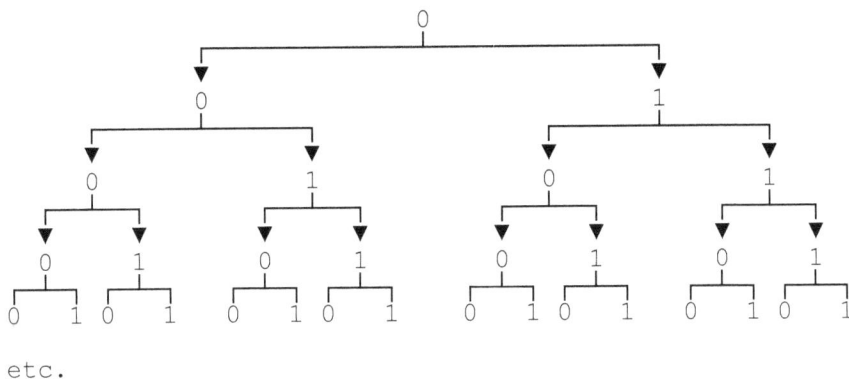

etc.

    Well, I hope it's obvious that this structure, extended indefinitely, can be understood as representing every real of the form

$$0.t_1 t_2 t_3 t_4 \ldots$$

where t_n ($n = 1, 2, \ldots$) is the nth bit, 0 or 1, in the fractional part, expressed in binary, of the real in question: Just start at the top of the tree and follow the arrows, branching to the left or the right according as the next bit of the fraction in question is 0 or 1. In other words, the tree represents all of the real numbers in the interval [0...1]. Terminology:

a. The tree consists of a set of *nodes* and a set of *links* (and those links form *paths*—see point e. below).

b. The top or so called "root" node corresponds to the initial 0 that precedes the (binary) radix point in all of the (binary) real numbers that we're interested in—which is to say, the real numbers in the interval [0...1].

c. Every other node corresponds to some specific bit in the (binary) fractional portion of some specific one of those real numbers.

d. Every link is the edge, or connection, from some specific node to one of the two immediately adjacent nodes on a downward path through the tree.

e. The sequence of links from the top of the tree down to any particular node constitutes the path to that node. *Note:* Here and throughout what follows I use the term *sequence* not as elsewhere in this essay to mean a sequence of real numbers specifically, but rather just in its usual informal sense.

    Note that the tree is a binary tree specifically, in the sense that every node has exactly two links emanating from it.

    Now let me explain how this tree structure can be used to count the reals. I'll explain the basic idea first, then deal with various points and/or objections that might occur to you in connection with that preliminary explanation.

The Basic Idea

- First, let's agree to refer the root node as the zeroth *level* of the tree.

- Next, let the set of links to nodes immediately reachable from the root node, together with those immediately reachable nodes, constitute the next (i.e., first) level of the tree. There are $2^1 = 2$ links at that level; number them 1 and 2, left to right. (There are also two nodes at that level accordingly, of course.)

- Similarly, let the set of links to nodes immediately reachable from nodes at the first level, together with those immediately reachable nodes, constitute the next (i.e., second) level of the tree. There are $2^2 = 4$ links at that level; number them 3, 4, 5, and 6, left to right. (There are also four nodes at that level accordingly.)

- And so on. That is, for all $n > 0$, let the set of links to nodes immediately reachable from nodes at the $(n-1)$st level, together with those immediately reachable nodes, constitute the nth level of the tree. There are 2^n links at that level; number them 2^n-1, 2^n, ..., $2^{n+1}-2$, left to right. (There are also 2^n nodes at that level accordingly.) Note too that adding one to $2^{n+1}-2$ yields $2^{n+1}-1$, which is the first number to be assigned at the next or $(n+1)$st level, and so there are no gaps in the numbering.[3] Nor are there any duplicates.

The foregoing scheme, as n goes to infinity, assigns a unique natural number to every link and a unique link to every natural number; in other words, it defines a one to one onto mapping between the links and the natural numbers.[4] So the links are countable.

Now let's think about paths (recall that a path is the sequence of links from the root down to a particular node). Clearly, the number of distinct paths, nP say, to nodes at the nth level ($n > 0$) is the same as the number of nodes at that level—they're both equal to 2^n. (Also, each such path represents some specific real to exactly n bits of precision, without rounding.) As n goes to infinity, therefore, the total number of paths—and hence the total number of reals in the interval [0...1], because every such real is represented by some such path—also both go to infinity.

Now, the number of distinct links, nL say, involved in paths to nodes at the nth level is $2^1+2^2+2^3+ ... +2^n$. So for any given n there are $nL-nP = 2^1+2^2+2^3+ ... +2^{n-1}$ more links than paths—from which it follows that since the links as n goes to infinity are countable, so the paths as n goes to infinity are countable as well, and hence the reals in the interval [0...1] are countable also.

---

[3] Sayan's scheme for numbering links is different from mine, but not in any important respect.

[4] Since for any given node there's exactly one immediate link to that node on a downward path through the tree (and vice versa), we could conclude with equal justification that it defines a one to one onto mapping between the *nodes* and the natural numbers.

But what about the set of *all* reals? Well, all we have to do here is consider a set of trees, one for each of the intervals [0...1], [1...2], [2...3], and so on, where each tree in the set is analogous to the one discussed in detail above (though for clarity I've shown the various boundary values—somewhat inconsistently!—in decimal instead of binary). Clearly:

a. Each such tree represents a countable set, viz., the set of all reals in the corresponding interval;

b. The complete set of all such trees taken together serves to represent the reals in their entirety;[5] and

c. The number of trees in that set is countable as well.

Thus, since "countable times countable is countable,"[6] it follows that the reals in their entirety are countable. QED.

Let me now spell out for the record a couple of points arising from the foregoing (the points in question have probably already occurred to you, but I want to state them explicitly anyway):

■ First, recall the following text from the end of the previous section:

> [It] doesn't follow ... that we can't enumerate, or count, the reals in some other way—and in fact we can, as the section immediately following demonstrates.

Well, I hope it's clear that the scheme presented above for numbering the links in the tree for the interval [0...1] constitutes, precisely, a method for counting the reals—the reals in that interval, at any rate.

■ Second, the process of building that tree is logically equivalent to exhaustively evaluating a certain generating function that yields all of the reals in that interval. In this connection, let me remind you of the following remarks from the section "Some Questions" in Part II of this essay:

> [If a sequence] is infinite, then the *only* way it can it be defined is by means of some rule, or in other words some algorithm or "generating function" f that gives the nth element of the sequence for all positive integers n.

---

[5] Recall that for simplicity I'm ignoring reals less than zero. However, let me note for the record that—as you might have realized for yourself—certain reals (to be specific, those that happen to be positive integers) will be represented twice in the proposed complete set of trees. For example, the positive integer 1 will be represented by both the rightmost infinite path in the tree for [0...1] and the leftmost infinite path in the tree for [1...2]. But of course this state of affairs has no effect on questions of countability. See the subsection "Rational Numbers," later in the present section, for further discussion of such matters.

[6] Another theorem whose proof I'll leave as an exercise. (*Hint:* Cantor's proof that the rationals are countable—which I'll be touching on, briefly, in Appendix C to this part of the essay—can be adapted for the purpose.)

Irrational Numbers

Now let me turn as promised to various points and/or objections that might have occurred to you in connection with the foregoing argument. The first is this: Doesn't the tree as I've described it represent only rational numbers ("rationals"), not all possible reals? As one reviewer put it: "I can keep on descending levels but I don't see how I [can] ever arrive at an irrational number."[7]

Now, it's certainly true that for all finite n the reals represented by the 2^n paths to nodes at the nth level are all exactly n bits long—I mean, they're all of the form $0.t_1t_2...t_n$—and hence they're all rationals, by definition. Indeed, it's true that an irrational number can't be represented precisely by n bits for any finite n—to do so (to represent it precisely, that is) requires an infinite number of bits, also by definition. But an infinite number of bits is exactly what the tree structure gives us! In fact, I'd like to turn the question around: I challenge you to show me an irrational number that the tree *doesn't* represent.

Let me put the point another way. How are irrational numbers defined, anyway? *Answer:* By means of some infinite series. For example, the number π satisfies the following identity ("Leibniz's formula")—

$$\frac{\pi}{4} = \frac{1}{1} - \frac{1}{3} + \frac{1}{5} - \frac{1}{7} + \frac{1}{9} \ldots$$

—from which it follows immediately that π (which is certainly irrational, and in fact transcendental) can be defined thus:

$$\pi \stackrel{\text{def}}{=} \frac{4}{1} + \frac{(-4)}{3} + \frac{4}{5} + \frac{(-4)}{7} + \frac{4}{9} \ldots$$

Evaluating the series on the right here to any finite number of elements obviously yields a rational result—and yet the series overall denotes an irrational number. In other words, just to spell the point out: To reject the idea that an infinite path can represent an irrational number is equivalent to rejecting the idea that an infinite series can represent an irrational number.

At the risk of beating a dead horse, let me pursue this point just a moment longer. To repeat, an irrational number is an infinite series *by definition*.[8] For example, using decimal notation, the number π simply *is* the following series—

$$\pi \stackrel{\text{def}}{=} x_0 + x_1 + x_2 + x_3 + x_4 + x_5 + \ldots$$

—where the individual elements (the x's) are as follows:

---

[7] I note in passing that an exactly analogous argument could be made in connection with a Cantor list. That is, if the objection were valid (it isn't), it would have to apply to Cantor lists as well—i.e., such a list too would have to consist of rationals only.

[8] Note the implication: The only way an irrational number can be completely defined, or completely specified, is by providing *an infinite amount of information.*

$$
\begin{aligned}
x_0 &= 3.0 \\
x_1 &= 0.1 \\
x_2 &= 0.04 \\
x_3 &= 0.001 \\
x_4 &= 0.0005 \\
x_5 &= 0.00009 \\
&\ldots \text{ and so on}
\end{aligned}
$$

For any finite n, the sum of the first n of these x's is always a rational number; however, the limit of that sum as n goes to infinity is irrational (in fact it's transcendental, as I've said). And yet the question "When does the sum switch from being rational to irrational?" is meaningless. The sum is irrational only when it's complete (i.e., when n goes to infinity).

More on Trees vs. Lists

By now it might have occurred to you to wonder, if the reals are truly countable (as Sayan and McGoveran and I, and many others, now claim), then why can't they be spelled out, at least conceptually, in the form of a Cantor list after all? That is, why doesn't countable imply listable? As I said in footnote 2 to this part of the essay, "Cantor listable" does imply countable, but the converse is false. But why?

Well, I don't think I can do better here than simply quote some remarks from Sayan's paper (though I've paraphrased them somewhat to make them fit more neatly into the present context, and I've added some boldface):

Some might argue that if the reals are countable, then it must be possible in principle to list them all in array form (i.e., in the form of a Cantor list). But that takes us back to the problem discussed earlier—viz., the conflict between the diagonal and the entry in the list for the antidiagonal. In other words, the fact that the reals are countable does *not* imply that they can be listed in array form. To say the reals are countable is simply to say they can be put into one to one correspondence with the natural numbers. **But an array or Cantor list involves something more than just such a correspondence—it involves the notion of a diagonal, which is something that has no counterpart in the pure notion of a one to one correspondence as such.**

Aside: I think what Sayan here refers to as just a one to one correspondence (i.e., between the reals and the natural numbers) might better be described as an *isomorphism*. An isomorphism is a one to one correspondence with the additional property that it "preserves structure." For example, the one to one correspondence between reals and their logarithms is an isomorphism, because the product of any two reals corresponds to the sum of their logarithms and vice versa. But when we try to use a Cantor list to represent a correspondence between natural numbers and reals, the "reals" side of the correspondence involves the concept of a diagonal, which is something that has no counterpart on the "natural numbers" side. *End of aside.*

Sayan then goes on to add the following (though again I've paraphrased the original text somewhat):

Hence, to assert that if the reals and the natural numbers are both countable then we must be able to set up a Cantor list pairing them off is unwarranted. The appropriate way to arrange the reals is as a set of trees [*which could in principle all be combined into a single tree if desired*], not as an array. Cantor's starting point (listing all the reals in an array) was mistaken.

McGoveran's Procedure

In his paper "Diagonalization," which I've mentioned several times in this essay already, David McGoveran describes another procedural approach to showing the reals are countable. The procedure in question works by building a structure logically equivalent to Sayan's tree structure, assigning natural numbers to nodes (or node equivalents, rather) as it goes, and thereby explicitly demonstrating countability. Here's one way to explain how that procedure works:

a. Construct a list containing all possible bit strings of length one:

```
0
1
```

Call this list L, and number the strings so listed 1 to 2 from top to bottom.

b. For all e, where e ("element") is a string in L, do 1.-3. below:

1. Make a copy f of e.

2. Extend e and f by appending a trailing 0 to e and a trailing 1 to f.

3. Revise L by (a) replacing e by that extended version of e and (b) inserting the extended version of f immediately following that extended version of e.

Number the strings in this new version of L consecutively from 1.[9]
    Note: The first time Step b. is executed, the new version of L looks like this:

```
0 0
0 1
1 0
1 1
```

The strings in this version of L are numbered 1 to 4 from top to bottom.

c. Repeat Step b. "forever."

---

[9] Actually McGoveran's numbering scheme doesn't start from 1 again at each step—rather, it continues from $m+1$, where m is the number last assigned on the previous step—but starting from 1 at each step might make it more directly obvious that the reals in their entirety are countable. (As a matter of fact McGoveran's own numbering scheme is essentially identical to the one I described earlier for numbering the nodes in Sayan's tree.)

Note: The first time Step b is repeated, the new version of L looks like this:

```
0 0 0
0 0 1
0 1 0
0 1 1
1 0 0
1 0 1
1 1 0
1 1 1
```

The strings in this version of L are numbered 1 to 8 from top to bottom.

Observe now that:

- After executing Step b. of the foregoing procedure $n-1$ times ($n \geq 1$), the list L will contain 2^n strings, and those strings will be numbered from 1 to 2^n, inclusive; in fact, L will be "complete" at this point, in the sense that it contains all possible strings of length n bits. (Equivalently, it contains the binary representation, to exactly n bits of precision without rounding, of the fractional portion of every number in the interval [0...1].)

- What's more, all of the foregoing properties continue to hold as n goes to infinity.

- Finally, it should be clear that at every stage of its construction the list L is logically equivalent to Sayan's tree structure, down as far as the level corresponding to that stage of construction. In other words, McGoveran's list and Sayan's tree are isomorphic to one another.[10]

Rational Numbers

Recall now the following text (repeated from Part I of this essay, but somewhat edited and abbreviated here):

> Certain reals have two distinct decimal representations; for example, the decimal strings 0.10000... and 0.09999... both denote the real "one tenth." So do we need to worry about this state of affairs? The short answer is "No, we don't—at least, not much" ... But I don't want to deal with the issue here in detail here; I'll come back and discuss it in detail in Part III of this essay.

Well, now it's time to get into that more detailed discussion. First let me switch to binary notation, as usual. Then it's clearly true that certain reals have two distinct binary representations. For example, the strings

---

[10] A slight oversimplification—McGoveran has asked me to note that the isomorphism isn't exact; more explicitly, that there are "subtle but important differences" between the two. However, the differences in question are beyond the scope of the present essay.

```
0.0110000000...   /* all subsequent bits 0 */
```

and

```
0.0101111111...   /* all subsequent bits 1 */
```

both represent the decimal number 3/8. Here's a slightly more precise characterization of the general situation:

> Every number in the interval (0...1) that has a binary representation ending in an infinite string of 0s has exactly one alternative binary representation ending in an infinite string of 1s, and vice versa.[11]

Of course, any number that has two distinct representations in the manner just described must be a rational number specifically, by definition. Let's agree to refer to such numbers for brevity as *DR rationals* (DR for doubly represented).[12] Then the tree structure we've been talking about will contain two distinct paths—and hence, in effect, two distinct representations—for each DR rational in the interval (0...1). *But it doesn't matter.* What I said before was this, more or less:

> As n goes to infinity, the total number of paths—and hence the total number of reals, because every such real is represented by some path—also both go to infinity.

What I carefully didn't say was that every real is represented by some *unique* path—and in fact we now know that, *very* loosely speaking, there are "more" paths than reals. (To state the matter more precisely, each real in (0...1) is represented by either one path or two—two if the real in question is a DR rational, one otherwise. See the aside below for further discussion.) But if there are more paths than reals, and if the paths are countable (which they are), then it follows that the reals must be countable too; in fact, they must be "even more" countable, as it were, than the paths are.

However, the foregoing discussion does point up a logical difference that in practice we often overlook: namely, the logical difference between

---

[11] Note the change here from the closed / closed interval [0...1] to the open / open interval (0...1), with the consequence that what I called that "slightly more precise characterization of the situation" doesn't apply to the boundary values 0 and 1. This change isn't really all that important in the overall scheme of things—I mean, it doesn't significantly affect the discussion—but I have two reasons for making it, one for each of the two boundary values. First, the rational 0, unlike all other rationals with a binary representation that ends in an infinite string of 0s, *doesn't* have an alternative binary representation that ends in an infinite string of 1s. Second, the rational 1 does have two distinct binary representations (viz., 1.00000... and 0.11111...), but only the second of these appears in the tree structure we've been talking about, viz., the one that represents the reals in the closed interval [0...1].

    PS: If (like one of my reviewers!) you're not convinced that certain rationals do have two distinct representations—and by the way, such rationals do always exist, regardless of whether the radix is binary or decimal or something else again—then you might find the Wikipedia article titled "0.999..." helpful. See also Appendix A to the present part of this essay.

[12] I toyed with the idea of calling them Jekyll and Hyde numbers, since we're talking about a situation in which there's really just one entity involved, but the entity in question displays what we might call two distinct personas. But I decided not to, because "Jekyll and Hyde" might suggest that one of those personas is somehow to be preferred over the other, which isn't the case (not logically, at any rate).

a. Numbers as such, on the one hand, and

b. Representations of such numbers, on the other.

Of course, I've touched on this issue in this essay several times already, but now it's time to get more specific. By way of example, consider the number "one half." That number is indeed a number, of course, but it can be represented as either 0.50000... or 0.49999... in decimal, or as either 0.10000... or 0.01111... in binary. Thus, the (infinite) paths in the tree structure we've been discussing don't correspond one to one with numbers, they correspond one to one with *representations*: specifically, binary representations. (Though they do also correspond one to one with numbers as such, so long as we restrict our attention to just those real numbers that don't happen to be DR rationals. Again, see the aside below.)

What about Cantor lists? Well, it's obviously the case that such lists too contain representations of numbers, not numbers as such, and hence that such lists will thereby contain two entries for every DR rational. And even if they didn't—i.e., even if some scheme were in effect according to which such numbers had only one entry in the list after all—the diagonalization process would still produce different results depending on which of those representations actually appeared. This state of affairs makes it clear that, at the very least, diagonalization has to do with representations of numbers and not with numbers as such. [13]

> *Aside:* The fact that there's a logical difference between numbers and representations does cause a slight complication with respect to the correspondence between reals and paths (where by "paths" I mean, more specifically, the paths in Sayan's tree that represent the reals in [0...1] to infinite precision, not the finite truncated versions of those paths). It's true that every such path corresponds to exactly one real. It's also true that every real r in [0...1]—or in (0...1), a fortiori—corresponds to exactly one such path, so long as r isn't a DR rational; but if r does happen to be a DR rational, then there are two such paths. So:
>
> ■ If r isn't a DR rational, let p be the sole path in the tree corresponding to r, and let t be the bit string represented by p.
>
> ■ If r is a DR rational, let $p1$ and $p2$ be the two (and only two) paths in the tree corresponding to r, and let $t1$ and $t2$ be the bit strings represented by $p1$ and $p2$, respectively. By definition, there'll be some finite n ($n \geq 0$) such that $t1$ and $t2$ match on their first n bits and differ ("for the first time," as it

---

[13] Not only does the diagonalization argument have to do with representations, but in order to be well defined and complete (if "completeness" even makes sense in this context) it clearly needs to include a proper treatment of the DR rationals—which is an interesting observation in itself, since few if any of the published accounts I've seen do include any such treatment. Note in particular that, absent any such proper treatment, the entry that diagonalization identifies as missing from some specific Cantor list might just be one of the two representations for some DR rational for which the other representation isn't missing at all!

were) on their $(n+1)$st bit. Let t be, arbitrarily, whichever of $t1$ and $t2$ has a 0 in that $(n+1)$st position.

Now we can define a one to one onto correspondence, not between the set of reals r in [0...1] and the set of paths, but rather between the set of reals r in [0...1] and the set of all such bit strings t. But that set of bit strings is itself clearly countable, and so the desired result—that the reals in [0...1] are countable, and hence that the reals in (0...1) are also countable—still follows. *End of aside.*

For more on the logical difference between numbers and their representation, see Appendix A to this part of the essay.

The End (for Now)

My aim in the last few subsections has been to forestall, or attempt to forestall, various objections that might be raised in connection with Sayan's tree and the use of that tree in showing that the reals are countable after all. As I've found from experience, however, the arguments I've presented so far (i.e., in those subsections) are still not sufficient to silence the critics. But there are other issues, issues I haven't yet had a chance to discuss properly, that I personally regard as more pressing, and I really don't want to take the time to respond to those critics any further at this juncture. At the same time I do think it's important not to let their objections go unanswered, and so I've decided to defer the remainder of the debate to an appendix to this part of the essay (Appendix C). Obviously you can jump to that appendix right away if you like; however, what I'm going to do now is move on to consider those issues that I regard as more urgent. Specifically, I want to revisit Cantor's Theorem—his power set theorem, that is—and examine it carefully in the light of what we've learned so far.

CANTOR'S POWER SET THEOREM REVISITED

I discussed Cantor's power set theorem—more usually known just as "Cantor's Theorem," unqualified—in Part I of this essay. Here just to remind you is a formal statement of that theorem:

> **Cantor's Theorem:** Let s be a set and let S be the power set of s (i.e., the set whose elements are all of the subsets of s). Then the cardinality $|S|$ of S is strictly greater than the cardinality $|s|$ of s.

Of course, this result is well known (and noncontroversial) if s is finite—in that case, the cardinality of S is simply 2^n, where n is the cardinality of s, and as I showed in Part I of this essay, 2^n is strictly greater than n for all $n \geq 0$. But Cantor's Theorem is supposed to apply in particular to the case where s is infinite. Now, I did discuss Cantor's proof of this latter case in Part I of this essay; however, I also suggested, in a footnote to that discussion, that the proof in question could be seen as yet another application of Cantor's diagonalization scheme and might therefore be regarded with suspicion. It's time to take a closer look. In what follows, I'll show, first, that Cantor's proof is indeed basically just diagonalization once again; second, and more particularly, that Sayan's "non number" objection therefore applies to it.

Note first of all that Cantor's Theorem implies in particular that if N is the (infinite) set of natural numbers, then the cardinality of the power set of N is strictly greater than that of N itself, and hence that that particular power set is uncountable. So let's focus on that specific case; i.e., let sets s and S be the set N of natural numbers and the power set of that set N, respectively.[14] Now imagine an infinite array—in effect a Cantor list—constructed as follows:

- Let there be one row for each element of s (i.e., for each natural number), and no other rows.

- Let there be one column for each element of S (i.e., for each set containing nothing but natural numbers), and no other columns.

- Let the element at the intersection of row i and column j be 1 if the natural number corresponding to row i—which is of course just i itself—is an element of the set corresponding to column j, and 0 otherwise.

Here are the first few entries in such an array—such a Cantor list, if you prefer—with the diagonal highlighted:

	{2,3,7,9}	{4,6}	{1,3,15,60}	{ }	{2,5,68}	{7}	{8,15,36}	...
1.	*0*	0	1	0	0	0	0	...
2.	1	*0*	0	0	1	0	0	...
3.	1	0	*1*	0	0	0	0	...
4.	0	1	0	*0*	0	0	0	...
5.	0	0	0	0	*1*	0	0	...
6.	0	1	0	0	0	*0*	0	...
7.	1	0	0	0	0	1	*0*	...

etc.

As you can see, the diagonal d in this example is

0　0　1　0　1　0　0 ...

So the inverse, or antidiagonal, d' is

1　1　0　1　0　1　1 ...

Next, define the set sd to consist of just those natural numbers—equivalently, those row numbers—for which d contains a 1:

$$sd \overset{\text{def}}{=} \{\, 3\, ,\, 5\, ,\, \ldots\, \}$$

---

[14] The example that follows is again based (albeit rather loosely) on one from Sayan's paper.

Analogously, define the *inverse set sd'* to consist of just those natural numbers (i.e., row numbers) for which d' contains a 1:

$$sd' \overset{\text{def}}{=} \{\ 1\ ,\ 2\ ,\ 4\ ,\ 6\ ,\ 7\ ,\ \dots\ \}$$

Now, sd and sd' are clearly both subsets of N, the set of natural numbers— i.e., they're both elements of S—and so they must each correspond to some column of the array. The column for sd (though it's not shown in that portion of the array depicted on the previous page) reads from top to bottom as follows:

```
0   0   1   0   1   0   0 ...
```

And the column for sd' reads from top to bottom as follows:

```
1   1   0   1   0   1   1 ...
```

Note carefully that these two columns are identical in content to the diagonal d and the antidiagonal d', respectively—as indeed they must be, because of the way they're defined.

Now let's focus on the set sd'. To repeat, that set is clearly an element of S, and so it must correspond to some column of the array. *But it can't.* Why not? Well:

- Suppose it corresponds to the nth column.

- By definition, then, the diagonal d will intersect that column at the nth bit position.

- So if the nth bit of that column is 0, the nth bit of the diagonal d must also be 0.

- So the nth bit of the inverse (antidiagonal) d' of d must be 1.

- But the nth column (i.e., the column for sd') is, by definition, identical in content to the antidiagonal d'. So the nth bit of that column must be 1.

- In brief, therefore: If the nth bit of the column for sd' is 0, then it must be 1; and by a precisely analogous argument, if it's 1, then it must be 0. Contradiction!

- So the column for sd' can't appear after all as the nth column for any n; in other words, it can't appear in the array at all.

Now, Cantor concludes from the foregoing argument—or from his version of the foregoing argument, I should rather say, but then his version is logically equivalent to the version I've just presented—that there must be strictly more elements in the power set of the natural numbers than there are natural numbers, and hence that the power set of the natural numbers must be uncountable. But that argument of Cantor's suffers from at least one flaw, and arguably from two different ones:

- First, it's a diagonalization argument, and is thus subject to the criticisms of all such arguments identified in Part I of this essay.

- Second, even if we ignore those criticisms, the assumption that the array has a column for every element of S implies that diagonalization fails, precisely because it involves what Sayan calls "non numbers."

So Cantor's "proof" (i.e., the "proof" of Cantor's Theorem) isn't valid. But is what that proof purports to prove valid nonetheless? No, it isn't. To demonstrate this fact, Sayan returns to the tree structure discussed in the section "Uncountability of the Reals Revisited"—which represented all of the reals in the interval [0...1], remember—and argues as follows.

First, recall that every link in that tree can be defined to correspond to, or in effect be labeled with, a unique natural number. (Just to remind you, the links at the nth level, $n \geq 1$, were numbered 2^n-1, 2^n, ..., $2^{n+1}-2$, left to right.) So any arbitrary set sL of links—which is to say, any arbitrary element of the power set of the complete set of links in the tree—can be defined to correspond to a bit string that has a 1 in the nth bit position if and only if link n is an element of sL. For example:

```
If the set of links sL is:          Then the bit string is:

    {1}                              1000000000000000...
    {2,3,8}                          0110000100000000...
    {3,6,7,...}                      0010011000000000...
    {5,10,15,16,17,...}              0000100001000111...
    {every link}                     1111111111111111...
    {}                               0000000000000000...
```

This correspondence is clearly one to one and onto, because for each distinct set of links sL there's a unique bit string and vice versa. And given also that—modulo the discussion in the aside immediately following—there's also a one to one onto correspondence between those bit strings and the reals in the interval [0...1], it follows that for each distinct set of links sL there's a unique real number in the interval [0...1] and vice versa. So the power set of *all* links in the tree has the same cardinality as the set of all reals in the interval [0...1]. (In other words, using notation and terminology to be explained in just a few moments, we have $|PTL| = |ZO|$.)

Aside: Of course, here once again we do need to deal with the fact that certain reals—specifically, the ones I referred to a few pages back as "DR rationals"— have two distinct bit string representations. For example, the bit strings 10000... and 01111... both represent the real "one half," and that real thus corresponds to both (a) the set of links containing just link 1 and (b) the set of links containing all links apart from link 1. In a sense, therefore, there are "more" sets of links than there are reals. To repeat, however, it's precisely those reals that happen to be DR rationals that account for the discrepancy; and since the rationals in their entirety are countable (and the DR rationals in particular are therefore countable a fortiori), it follows that the discrepancy in question has no material effect on the argument overall. We can safely ignore that discrepancy, therefore, and for simplicity I'll do so from this point forward. *End of aside.*

It's convenient to introduce some abbreviations. To be specific:

■ Let *TL* ("total links") denote the set of all links in the tree—the tree, that is, that represents the reals in [0...1]—and let *PTL* denote the power set of *TL*.

■ Similarly, let *ZO* ("zero to one") denote the set of all reals in the interval [0...1], and let *PZO* denote the power set of *ZO*.

■ Let $|TL|$, $|PTL|$, $|ZO|$, and $|PZO|$ denote the cardinalities of *TL*, *PTL*, *ZO*, and *PZO*, respectively.[15]

Now, we've already seen that sets *ZO* and *TL* are both countable; thus $|ZO| = |TL|$. Now, there's a theorem of set theory that says that if two sets have the same cardinality (n, say), then their power sets also have the same cardinality (N, say).[16] So $|PZO| = |PTL|$. But we saw a few moments ago that $|PTL| = |ZO|$, and so we have $|PZO| = |ZO|$. In other words, the power set of *ZO* has the same cardinality as *ZO* itself; thus, since *ZO* is countable, it follows that its power set *PZO* is countable as well.

So much for the reals in [0...1]; but what about the set of *all* reals in their entirety? Well, we saw in the section "Uncountability of the Reals Revisited" that the set of all reals is countable, and hence has cardinality the same as that of *TL*. So the power set of all reals has cardinality $|PTL| = |ZO|$, and is therefore countable in its turn. QED.

As a kind of postscript to the foregoing, let me remind you of the following (paraphrased from Part I of this essay):

[What the arguments we've been discussing refute] isn't just Cantor's specific "proof" that the reals are uncountable—it's a refutation of the logic of diagonalization in general. That is, *any* result that's derived from diagonalization in some shape or form is suspect, or at best unproven.

Well, we now know in particular that diagonalization, albeit in somewhat concealed form, is precisely what Cantor used to obtain the result known as Cantor's Theorem: viz., that the cardinality of the power set of a given infinite set is strictly greater than that of the set in question. So perhaps it isn't all that surprising that the result in question turns out not to be valid. At the same time, the fact that it does (turn out not to be valid, I mean) gives still more support to the claim that *any* result obtained via diagonalization needs to be taken with a very large pinch of salt.

---

[15] $|TL|$ in particular is the limit as n goes to infinity of what I referred to earlier as nL, the number of links involved in paths from the root node to nodes at the nth level.

[16] Of course, this result is obvious if the sets in question are finite, because then N is 2^n in both cases.

APPENDIX A: REAL NUMBERS AND THEIR REPRESENTATION

The aim of this appendix is to elaborate briefly on the logical difference, already touched on at several points earlier in this essay, between (a) real numbers as such, on the one hand, and (b) the representation of such numbers using some particular base or radix, on the other.

For simplicity once again, throughout this appendix I'll consider just the fractional part of the numbers we're talking about; that is, I'll ignore the radix point and everything before it—barring explicit statements to the contrary, of course—and consider just strings of digits, and thus (in effect) interpret those strings of digits as representing real numbers r in the interval [0...1]. Then:

- I hope it's clear right away that the logical difference we're talking about is indeed—forgive me—real, if only because a given real number r can be represented using any number of different radixes (decimal, binary, etc.).[17] In other words, there's a one to many correspondence between real numbers and their representations.

- As mentioned at several points earlier in this essay, Cantor's lists and Sayan's trees don't contain numbers as such—instead, they contain representations of numbers using some specific radix (usually binary, for reasons of both definiteness and simplicity).

- Because of the previous point, care is needed in deriving results from those lists and trees that do truly apply to numbers as such and not merely to representations (to binary representations in particular). And I venture to suggest that, in the case of Cantor's lists though not in the case of Sayan's trees, sufficient care has sometimes *not* been taken in the past, with the consequence that certain implications have been drawn from those lists that can be seen on closer inspection not to be valid.

- With respect to representations specifically: The representations in question are always strings of infinite length, at least in principle, even if in some cases the digits after some specific point in the string are either all equal to 0 or all equal to $R-1$, where R is the pertinent radix. Here are a couple of decimal examples:

```
1250000... /* 0s ad infinitum after third digit */

1249999... /* 9s ad infinitum after third digit */
```

---

[17] Technically, the term *radix* refers to the integer that serves as a basis for those representations, and so it makes no sense to say that, e.g., "decimal" is a radix. Rather, decimal is the representation we get when we use ten as the radix; likewise, binary is the representation we get when we use two as the radix, and so on. But I don't think it'll cause any confusion if I continue to refer to decimal, binary, etc., as radixes per se.

■ The strings just referred to—i.e., those in which the digits after some specific point are either all equal to 0 or all equal to $R-1$, where R is the radix—all correspond by definition to rational numbers ("rationals") specifically. In decimal, for example, the strings shown above both represent the rational 1/8 (one eighth). Let's agree to refer to such rationals as *rationals of the first kind*.

■ Given any particular radix, however, there'll always be some rationals that can't be represented in the foregoing manner. In decimal, for example, the rational 1/7 (one seventh) begins thus—

```
142857142857...
```

—and that sequence of digits 142857 goes on to repeat, or recur, an infinite number of times ("indefinitely"). Let's agree to refer to such rationals as *rationals of the second kind*.

Note: In general, a representation is said to *repeat* (or *be repeating, or be periodic*) if it eventually reaches a point after which the same sequence of digits, not all equal to 0 and not all equal to $R-1$ where R is the radix, recurs indefinitely. For example, the decimal representation of "one seventh" repeats (i.e., is periodic), as we've just seen. For another example, the binary representation of "one sixth" takes the form 00101010... and thus repeats (or is periodic) because, after the two initial 0s, the sequence of digits 10 recurs indefinitely.

Of course, rationals of the first kind are what I earlier called "DR rationals" (DR for doubly represented), precisely because they have the property that they can be represented, in the specific radix we happen to be using, by two distinct strings. For example, the two strings shown above—

```
1250000...
1249999...
```

—both represent the rational 1/8 in decimal (the ellipsis in the first of these strings stands for an infinite string of trailing 0s, that in the second for an infinite string of trailing 9s). Likewise, in binary, that same "rational of the first kind" 1/8 can be represented by either of these strings:

```
0010000...
0001111...
```

The ellipsis in the first of these strings stands for an infinite string of trailing 0s, that in the second for an infinite string of trailing 1s.

To recap: Let r be a real number and let R be a radix. Then:

■ If r is irrational, then no matter what radix R we happen to be using the precise string representation of r is always infinitely long and nonrepeating.

In decimal, for example, the precise string representation of the fractional portion of √2 (which is irrational) begins

```
41421356...
```

and continues on to infinity without ever becoming periodic.

■ If and only if

    a.   The number r is rational and can be represented precisely, using radix R, as a finite string of significant digits $f1$ followed by an infinite number of trailing zeros, then

    b.   It can also be represented precisely using radix R by a slightly different finite string of significant digits $f2$ followed by an infinite number of trailing digits d, where $d = R-1$. For example, if R is ten (decimal), d is 9; if R is two (binary), d is 1.

    The two finite strings of digits in question, $f1$ and $f2$, differ only in the least significant position. Let the digits in that position be $d1$ for $f1$ and $d2$ for $f2$. Then $d2 = d1-1$. (The "one eighth" examples above illustrate.)

■ It's perfectly possible for a given rational r to have just one representation in one radix but two in another. For example, the rational 2/3 (two thirds) has just one representation in decimal—

```
6666666...
```

—but two in ternary (radix 3):

```
2000000...
```
```
0222222...
```

Thus, which rationals are of the first kind and which are of the second depends on what radix is being used.

■ As the foregoing bullet point shows, a given rational might be capable of precise representation by a finite number of significant digits using one radix but incapable of such representation using another. Here's another example: The rational 1/7 (one seventh) requires an infinite number of digits in decimal but just one, a solitary 1, in septenary (radix 7). By contrast, the precise representation of an irrational always requires an infinite number of digits, no matter what the radix.

    Now let rational number r and radix R be such that, using R, r has two distinct representations $r1$ and $r2$. Then $r1$ and $r2$ are members—in fact, the only members—of the same *equivalence class*, where the equivalence in question is "denotes the same number as, using R." For example, in decimal, the strings $r1 = 1250000...$ and $r2 = 1249999...$ are equivalent (both denoting as they do the same number, one eighth). Of course, those strings as such are distinct; thus, to say they're equal—

$$1250000..._{10} \quad = \quad 1249999..._{10}$$

(note the "10" suffixes, which make the radix explicit)—is strictly incorrect, even though we often do it. Rather, it would be more correct, or at any rate more accurate, to say they're *equivalent*—

$$1250000..._{10} \quad \equiv \quad 1249999..._{10}$$

(note the equivalence symbol "$\equiv$")—equivalent, to say it again, precisely in the sense that they both denote the same number.

APPENDIX B: UNFINISHED BUSINESS

This appendix consists of a number of lightly edited quotes, shown in italics, from the section "Unanswered Questions" in Part I of this essay, together with a little further commentary in each case (commentary, that is, that I wasn't in a position to provide previously).

Cantor's Theorem does seem to show there are an infinite number of "transfinite cardinals." Why? Because if the cardinality $|s|$ of an infinite set s is strictly less than the cardinality $|S|$ of its power set S, then $|S|$ in turn must be strictly less than the cardinality $|S'|$ of the power set S' of S, and so on ad infinitum.

Comment: Well, conceptually speaking there's certainly at least one infinity, viz., the infinity that's the cardinality of the set N of natural numbers, and we can call that one a "transfinite cardinal" if we like, and we can even label it "$\aleph_0$" if we like. [18] But as for there being an infinite number of such cardinals ($\aleph_0$, $\aleph_1$, $\aleph_2$, etc.) ... Well, we now know that "Cantor's Theorem" isn't a theorem at all, and it doesn't prove anything (at least, not if the cardinalities in question happen to be infinite); to be specific, it doesn't prove that the cardinality $|s|$ of an infinite set s is strictly less than the cardinality $|S|$ of its power set S. As a consequence, it also doesn't prove the existence of any transfinite cardinals other than $\aleph_0$. In fact, we now know that if some set has cardinality $\aleph_0$, then so does its power set; thus, we actually have *no reason whatsoever* to believe in the existence of any transfinite cardinals at all, apart from $\aleph_0$.

Given the foregoing state of affairs, I feel bound to add that to call $\aleph_0$ a "cardinal" as such is misleading anyway, because it suggests that infinity is just a number like one, or two, or a million, or a googol, or a googolplex. It's not. It's

---

[18] My use of that preliminary qualifying phrase "conceptually speaking" in this sentence is intended to convey the point that the infinity (the *sole* infinity!) I'm talking about here is only "potential," not "actual" or "completed." In other words—and more simply, and as I would personally much prefer to put it—the set N of natural numbers has no upper bound. For further explanation, see Appendix C to Part II of this essay.

a concept. It means, loosely, "bigger than any number you can think of."
Numbers as such are finite.

> *Aside:* You might recall me saying something similar to the foregoing once
> before (in Part II, Appendix C, to be precise). Here's the pertinent text, now
> slightly edited and abbreviated:
>
>> Those definitions make it possible to reason about what happens if some
>> operation is repeated "forever," without suggesting in any way that infinity is just
>> another number. It's not. It's a concept.
>
> Well, I apologize for the repetition, but the point is so important that I do think it
> bears repeating. *End of aside.*

*Set s is infinite if and only if it can be put into a strict one to one correspondence
with some proper subset p of itself. Observe, therefore, that this definition
implies that $|p| = |s|$. Of course, given that p is a proper subset of s, we might
have expected that $|p| < |s|$ ... The apparent contradiction involved here serves, I
think, to raise some obvious questions regarding the propriety of applying
comparison operators such as "<" to infinite comparands—which is what the
proof, or at any rate the conclusion, of Cantor's Theorem most certainly does.*

Comment: Well, I suppose we might say, somewhat vacuously, that if there's
only one transfinite cardinal (viz., $\aleph_0$), then all transfinite cardinals are equal to
one another. For example, if s and p are, say, the set of nonnegative integers
{0, 1, 2, 3, ...} and the set of squares of those integers {0, 1, 4, 9, ...},
respectively (and note in this example that p is indeed a proper subset of s, as
required), then those two sets can clearly be put into one to one correspondence
with each other—for every integer in the first set there's a unique square in the
second, for every square in the second set there's a unique nonnegative square
root in the first. So the cardinalities of those two sets s and p are both $\aleph_0$, and
they're equal. But if there's only one transfinite cardinal, then it can never be the
case that one such is strictly less than another, and so the operator "<" doesn't
make much sense. So I think we should explicitly recognize that the operator
"<" doesn't apply to—in fact, isn't even defined for—"transfinite cardinals."
Which brings me to my next point.

*The transfinite cardinal n1 is defined to be equal to the transfinite cardinal n2 if
and only if there exist sets s1 and s2 whose cardinalities $|s1|$ and $|s2|$ are n1 and
n2, respectively, and there exists a one to one onto mapping between s1 and s2 ...
The transfinite cardinal n1 is defined to be less than or equal to the transfinite
cardinal n2 if and only there exist sets s1 and s2 whose cardinalities $|s1|$ and $|s2|$
are n1 and n2, respectively, and s1 is a subset of s2 ... The transfinite cardinal n1
is defined to be less than the transfinite cardinal n2 if and only if $n1 \leq n2$ is true
and n1 = n2 is false.*

Comment: The first of these definitions (for "=") is still correct but again, if
there's only one transfinite cardinal, more or less vacuous. The second and third

definitions (for "≤" and "<", respectively) serve no useful purpose, and I believe they should be dropped.

Transfinite arithmetic is rather simple ... Let at least one of the cardinal numbers n1 and n2 be transfinite; then n1+n2 = n1×n2 = max (n1,n2).

Comment: If there's only one transfinite cardinal, $\aleph_0$, these definitions reduce to saying just $\aleph_0 + x = \aleph_0 \times x = \aleph_0$ for all x (even if x itself is $\aleph_0$). As for raising x to the power $\aleph_0$—well, presumably the result is again just $\aleph_0$ (again, even if x itself is $\aleph_0$).

Cantor tried, but failed, to prove the Continuum Hypothesis: viz., the hypothesis that there's no set whose cardinality C lies strictly between $\aleph_0$ and $\aleph_1$, and hence that $\aleph_i$ is indeed the "next" transfinite cardinal after $\aleph_0$.

Comment: Well, it's not surprising that Cantor failed here, since if there's no transfinite cardinal $\aleph_i$ that's distinct from $\aleph_0$, then the hypothesis is meaningless, and trying to prove or disprove it is a waste of time.

However, it's now known—I'm deliberately simplifying here, somewhat—that the Continuum Hypothesis is independent of the axioms of conventional set theory, meaning that either it or its negation can be assumed to be true without leading to any inconsistencies.

Comment: Nothing to add here!

Let propositions P1 and P2 be as follows:

  P1. There exists a transfinite cardinal C such that $\aleph_0 < C < \aleph_1$.

  P2. There doesn't exist a transfinite cardinal C such that $\aleph_0 < C < \aleph_1$.

Clearly, each of P1 and P2 is the negation of the other ... It clearly can't be the case that P1 and P2 both have the same truth value. Rather, if one is true, then the other must be false. But it also doesn't seem to make much sense to say that, e.g., P1 is true and P2 isn't, because the choice between them is apparently arbitrary. Is it really the case that the world could have "gone either way," as it were, on such a fundamental issue? ... Since there doesn't seem to be any good reason for preferring P1 over P2 or the other way around, it's very tempting to suggest that both propositions are meaningless, and that the whole idea of transfinite numbers is nothing but a chimera.

Comment: Here I stand by my original remarks.

APPENDIX C: CONTINUING THE DEBATE

I showed an early draft of the section "Uncountability of the Reals Revisited" from the body of this part of the essay—i.e., the section having to do with Sayan's representation of the real numbers in the form of a binary tree—to several mathematically inclined friends and friends of friends. Now, you'll recall that I attempted to forestall possible objections to Sayan's tree scheme by explicitly including in that section a number of subsections considering various aspects of that scheme in more depth. (The subsections in question—there were four of them—had the titles "Irrational Numbers," "More on Trees vs. Lists," "McGoveran's Procedure," and "Rational Numbers," respectively.) But despite this attempt on my part, I still received quite a few negative comments from one party in particular; in fact, his criticisms—the party concerned was male—together with my responses grew into a fairly lengthy back and forth between us. And here I fear I have to report failure ... I didn't succeed, through those responses, in changing my critic's mind at all. Nevertheless, it occurred to me that it could at least be useful to document the entire correspondence, inasmuch as it does at least air many of the pertinent arguments and counterarguments. So what follows is a tidied up and edited version of our exchanges, in the form of various comments from the critic and specific responses to those comments by myself.

Note: Naturally I sought the critic's permission to include this material in this essay. He gave that permission, but requested that I include a couple of paragraphs of his own giving his point of view. Those paragraphs appear at the very end of this appendix.

One last point by way of introduction: The material that follows does unavoidably repeat, here and there, certain of the arguments that have already been made at one point or another in the body of the essay. I apologize for this state of affairs, but once again I feel the arguments in question are sufficiently important to bear such repetition. If you feel otherwise, then I apologize again.

Critic: The nodes at the first level of Sayan's tree—or, more correctly, the paths to those nodes, but for simplicity let's stay with just "nodes" for the time being—clearly correspond to the binary numbers 0.0 and 0.1 (zero and one half), or in decimal fractional notation

    0/2   1/2

Similarly, the nodes at the second level correspond to 0.00, 0.01, 0.10, and 0.11, or in other words the fractions

    0/4   1/4   2/4   3/4

The nodes at the third level correspond to 0.000, 0.001, 0.010, 0.011, 0.100, 0.101, 0.110, and 0.111, or in other words the fractions

    0/8   1/8   2/8   3/8   4/8   5/8   6/8   7/8

And so on. So we see that:

> *Every node in the tree corresponds to a decimal "vulgar fraction"—i.e., a rational—of the form n/d, where the denominator d is a power of two (2, 4, 8, 16, etc.) and the numerator n is a nonnegative integer in the range 0, 1, ..., d–1.* [19]

Thus, there aren't any nodes for rationals such as $\frac{1}{3}$ (i.e., rationals where d isn't a power of two), nor are there nodes for irrationals such as $\frac{1}{2}\sqrt{2}$ or $\frac{1}{4}\pi$ (which can't be expressed in the form n/d in any radix at all). [20] So the tree certainly doesn't represent all of the reals in the specified interval, and Sayan's argument therefore fails.

Response: Actually the point the critic's making here doesn't have much to do with Sayan's tree as such; rather, it has to do with whatever radix we happen to be using, and thus of course with binary notation in particular. [21] The binary representation of *any* number—any number in the specified interval [0...1] in particular, of course, but actually any number at all—is basically just a string of bits (corresponding therefore to a specific path in some Sayan tree). Here by way of example is the binary or bit string representation of an arbitrarily chosen number in the interval [0...1] (integer portion "0" and radix point "." and trailing fractional digits all omitted for simplicity as usual):

1000110101110001

And yes, such a string by definition does always represent—and can only represent—some proper fraction of the form n/d where d is a power of two and $0 \leq n < d$. [22] Hence, it does indeed follow that rationals such as $\frac{1}{3}$ and irrationals such as $\frac{1}{2}\sqrt{2}$ or $\frac{1}{4}\pi$ can't be represented—at least, not precisely—by any such string at all, [23] *so long as the string in question is finite*. But the paths in Sayan's tree aren't finite, they're infinite! As a consequence, *every* real in the pertinent

---

[19] Of course, some of the numbers represented at every level after the first are the same as ones already represented at some previous level (e.g., 2/8 = 1/4). Perhaps a better way to put it is: The path that represents such a number at one level is subsumed by the extended version of that same path that represents that same number at the next level down. So it's not so much a matter of, e.g., the number "one quarter" being represented an infinite number of different times, but rather—as explained in a little more detail in footnote 23 below—a case of it being represented to an infinite number of different precisions. But in any case none of this has any material effect on the big picture.

[20] The multipliers $\frac{1}{2}$ and $\frac{1}{4}$ in these examples are introduced merely to ensure that the values in question are in range, as it were (i.e., they're contained in the interval [0...1]).

[21] It thus applies to Cantor's lists just as much as it does to Sayan's trees. In other words, if you think that what we have here is a valid objection to Sayan's tree proof (it isn't), then to be consistent you must agree that it's a valid objection to Cantor's diagonal proof as well.

[22] Let me stress the point that what we're talking about here is really just representation. Thus, we might equally well say the conventional representation of a number in the range [0...1] as a string of *decimal* digits—again omitting the initial "0." and radix point "." and trailing digits for simplicity—always represents, and can only represent, a number of the form n/d where d is a power of *ten* and $0 \leq n < d$. It's equally true, and equally irrelevant.

[23] However, they *can* be represented to any finite precision we like, of course. In fact, for all $n > 0$ the path to the nth node for a given real in Sayan's tree represents that real to exactly n bits of precision, without rounding.

interval—rationals such as ⅓ and irrationals such as ½√2 and ¼π all included— is represented by some such path after all.

At the risk of beating a dead horse, let me say a little more about the case of π in particular. Of course, the tree for the interval [0...1] has no path for π. However, a tree for the interval [3...4] will include the following path—

```
001001000011111101...
```

—which are the first few bits in the fractional part of the binary representation of π. Again, though, the path must be understood as going on forever, just as the decimal string

```
1415926...
```

(the first few digits in the fractional part of the decimal representation of π) must also be understood as going on forever.

If you're still not convinced, I challenge you to identify a specific real in the interval [0...1] that the tree for that interval *doesn't* represent. To put it another way: I claim that every possible infinite string of 0s and 1s appears as a path in the tree. If you don't agree, please identify one that doesn't.

Critic: In the numbering scheme described in the body of this part of the essay, the links, and therefore the nodes, at the nth level of Sayan's tree are numbered from 2^n-1 to $2^{n+1}-2$, thus:

```
1st level:    1 , 2
2nd level:    3 , 4 , 5 , 6
3rd level:    7 , 8 , 9 , 10 , 11 , 12 , 13 , 14
```

And so on. Each such set of node numbers has cardinality twice that of its immediate predecessor; in fact, the nth such set consists of the next 2^n integers following the last of those in that immediate predecessor. Thus, each such set clearly has a finite number of elements. Hence the set of all such sets, although infinite in cardinality itself, contains no set of infinite cardinality. I don't see how an infinite set to which the individual sets approach can be defined. Which integer would be its least element? I can't grapple with the idea that a path from the top of this tree can be of infinite length when none of the rows it crosses is infinite in length.

Response: Let me take these various points one at a time.

- "Each such set of node numbers has cardinality twice that of its immediate predecessor; in fact, the nth such set consists of the next 2^n integers following the last of those in that immediate predecessor." Correct.

- "Thus, each such set clearly has a finite number of elements." Well, it's true that for all $n > 0$, the nth such set has finite cardinality (viz., 2^n). As n goes to infinity, however, so does that cardinality 2^n.

- "Hence the set of all such sets, although infinite in cardinality itself, contains no set of infinite cardinality." No, this isn't correct. Once again using the standard notation $|s|$ to denote the cardinality of set s, a corrected version of the statement would be:

> Let S_n be the set containing just the first n of those individual sets, and let X_n be the element of S_n of greatest cardinality. (Actually X_n is of cardinality 2^n.) For all $n > 0$, $|S_n|$ and $|X_n|$ are both finite. But as n goes to infinity, so do both $|S_n|$ and $|X_n|$. (To say it again, $|X_n| = 2^n$, and 2^n certainly goes to infinity as n does.)

Thus, the critic was right about S_n but not about X_n.

- "I don't see how an infinite set to which the individual sets approach can be defined." Well, consider the following analogy. Let $T_n = 1+2+3+ \ldots +n$. For all $n > 0$, T_n is clearly finite; but as n goes to infinity, so does T_n. (We might say, loosely, that the *limit* of T_n as n goes to infinity is itself infinite. Or, and in fact better, we can say that as n goes to infinity T_n *doesn't have a limit*; in other words, it's unbounded.) But note carefully that the question "For what value of n does T_n switch from being finite to being infinite?" makes no sense—it's meaningless.

By the way, there are parallels, of a kind, between the foregoing example (i.e., defining T_n to be the sum $1+2+3+ \ldots +n$) and something I said in the body of this part of the essay regarding π: viz., that π is defined as the sum $3+0.1+0.04+0.001+ \ldots$ (etc.). For all n, the sum of the first n of these latter elements is a rational number (a rational number strictly less than π, as it happens); yet the sum as n goes to infinity is π, and that sum is irrational (and in fact transcendental). And to say "I don't see how an infinite set to which the individual sets approach can be defined" is akin to saying "I don't see how an irrational number to which the sum of terms approaches in the expansion of π can be defined." Of course, it's true in the π example that the irrational number in question is of finite size, but to pin it down precisely requires summing an infinite number of elements. Or to put it another way: The value of π is bounded, even though any precise representation of it requires an infinite number of terms ("summands").

To pursue the point just a moment longer: In a way, the critic is right to question the idea that the infinite set in question "can be defined." In fact it can't!—not precisely, anyway. That's what *infinite* means (or part of what it means, at any rate)—it's indefinite, and undefinable.

Infinity is hard to grasp. Crucially, though, *it's not a number*; loosely, it's something that's bigger than every number. Despite such conceptual difficulties, however, the key point about the infinite—for present purposes, at any rate—is this:

> *Mathematicians make the notion of the infinite manageable by means of the very carefully and precisely defined concept of a limit.*

I defined that concept—the concept of a limit, that is—in Appendix C to Part II of this essay. Let me also repeat the following from Part I (another of my critics speaking):

> Date is evidently missing an understanding and appreciation of the rigorous framework that mathematics has developed for dealing with the infinite, including the theory of infinite sequences and series and convergence thereof, which is generally regarded as one of its crowning achievements.

Well, I explained in Part I why I deny the specific charge here (viz., that I'm "missing that understanding and appreciation"). Au contraire, in fact: I most certainly do understand and appreciate that "rigorous framework"; moreover, I do strongly agree with and support the critic's "crowning achievement" remark.

■ "Which integer would be [the] least element [of the infinite set to which the individual sets approach]?" Not all infinite sets have a least element (some do, some don't). For example, the set of all positive rationals, or in other words the set of all rationals greater than zero, has no least element. (By contrast, the set of all *nonnegative* rationals, or in other words rationals greater than or equal to zero, does have such an element: viz., zero itself.) And I would say "the infinite set to which the individual sets approach" is a case in point—it has no least element.[24] After all, the elements of the nth such individual set are 2^n-1, ..., $2^{n+1}-2$, all of which become infinite when n does; thus, to say that one of these elements is "least" when n becomes infinite would be to suggest that one infinity might be less than another, a notion I've already rejected in Appendix B to this part of the essay.

■ "I can't grapple with the idea that a path from the top of this tree can be of infinite length when none of the rows it crosses is infinite in length." This comment needs a little clarification:

    a. By "row" I assume the critic means the set of links—or perhaps, but equivalently, the set of nodes—at some specific level of the tree.

    b. By "length" I assume he means the cardinality of that set.

But if these interpretations are correct, then the critic's claim that "none of the rows it crosses is infinite in length" is wrong. It's true that for all n, the length L_n of row n is finite; but as n goes to infinity, so does L_n.[25]

Critic: I accept the notion that a series can be convergent; for example, I agree that the sum

---

[24] It has no greatest element either.

[25] Oddly enough, the critic's very next objection, q.v., includes the words "But Sayan's tree doesn't converge!"—and from the critic's own explanation of what he means by this statement, it appears that he agrees that L_n does go to infinity after all (i.e., the rows do become infinite in length, as I've just said).

```
1/2 + 1/4 + 1/8 + ...
```

converges on the value one (unity) as the number of terms increases.[26] But we can't and don't calculate that result by performing an infinite number of additions; rather, we simply *define* it to be the value that the sum approaches, asymptotically, but never quite reaches. But Sayan's tree doesn't converge! There's no bottom row that can be defined as the limiting case. I accept that the *rationals* can all be represented by Sayan's tree: They correspond to paths that either terminate or converge. But I don't think even those can be counted using Sayan's method (I do know the method usually given for counting the rationals). I also accept that some irrational numbers (e.g., π) are defined by a convergent series, but Sayan claims that they can all be thus defined and I simply don't get it.

Response: Again I'll take the critic's objections one point at a time:

- "I accept the notion that a series can be convergent; for example, I agree that the sum

  ```
  1/2 + 1/4 + 1/8 + ...
  ```

  converges on the value 1 (unity) as the number of terms increases. But we can't and don't calculate that result by performing an infinite number of additions; rather, we simply *define* it to be the value that the sum approaches, asymptotically, but never quite reaches." Yes, such matters are what the mathematical notion of a limit is all about. For example, let the nth element of the series just shown be denoted x_n—so $x_n = 1/(2^n)$—and let the sum of the first n elements $x_1 + x_2 + \ldots + x_n$ be denoted T_n. Then the limit of T_n as n goes to infinity is 1, because for all $\varepsilon > 0$ there exists an integer N such that if $n > N$, then the difference between T_n and 1 is less than ε. Again, see Appendix C to Part II of this essay if you need to refresh your memory regarding the mathematical definition of what it means for something to have a limit, or limiting value.

- "But Sayan's tree doesn't converge! There's no bottom row that can be defined as the limiting case." I agree that the tree as such "doesn't converge," if by that wording the critic means simply that the rows grow in size without bound as we move further and further down the tree. Ultimately, in fact, the rows are infinitely long. But that's irrelevant. What's relevant is that every *path* converges. In fact, it converges in

---

[26] The critic's wording here is a trifle sloppy. Here's a more accurate (and more general) statement: Let S be an infinite series, and let the sum of the first n terms of S be S_n ($n > 0$). If S_n has a limit s as n goes to infinity, then that limit s is the limit of the series S, and we say, loosely, that S converges on s. As you'll see, the first of my responses to this particular comment by the critic elaborates on this point, slightly.

It's worth emphasizing that a series can be convergent in the sense just defined only if its terms have limiting value zero. Note that "only if," by the way—i.e., "only if," not "if and only if." Here's an example to illustrate the difference: The terms of the harmonic series $1 + 1/2 + 1/3 + 1/4 + \ldots$ certainly have zero as their limiting value, and yet the series diverges (i.e., it has no upper bound).

exactly the same way as the conventional (binary) representation of any real number in [0...1] converges: unsurprisingly, because such a path *is* the binary representation of such a number. As I said in footnote 23 to this part of the essay, for all n the path to the nth node for a given real—the nth node in that path, in other words—represents that real to n bits of precision, without rounding.

■ "I accept that the *rationals* can all be represented by Sayan's tree: They correspond to paths that either terminate or converge." I don't understand the critic's position here. Certain rationals—what I called in the body of this part of the essay "rationals of the second kind" (i.e., ones whose representation repeats)—can't be represented precisely in any finite number of bits. So if the critic is prepared to admit that the tree can represent such rationals nevertheless,[27] why isn't he prepared to admit that the same is true of irrationals, which likewise can't be represented precisely in any finite number of bits?

Now, if you think what follows is just repetition then I apologize, but I want to be sure my point is clear. Consider the rational number "one sixth." Ignoring the initial "0." as usual, here's the decimal representation:

```
166666666...
```

And here's the binary equivalent:

```
0010101010101010101...
```

And, of course, this latter string is identical to a certain path in Sayan's tree—which is hardly surprising, because that path is nothing more than just another way of depicting that string. Of course, both the path and the string (as well as the decimal analog 166666666...) do have to be understood as being infinitely long.[28] Thus, Sayan's tree can certainly represent all *rationals* in [0...1], not just those whose value in fractional notation have a denominator that's a power of two—despite the fact that these latter are the only ones represented in any finitely truncated version of that tree.

So if we can agree that "rationals of the second kind" are represented in the tree, why can't we agree that irrationals are represented in the tree as

---

[27] In fact the critic is contradicting himself here—earlier he said that Sayan's tree isn't capable of representing such rationals, but now he says it is.

By the way, that phrase "terminate or converge" is a little odd. "Terminate" presumably refers to the fact that some paths in the tree consist of a finite string of 0s and 1s, followed by nothing but 0s (and/or nothing but 1s, perhaps). But as for "converge"—well, *every* path converges, so I'm not quite sure what the critic is getting at here. Perhaps the reference is specifically to paths that correspond to rationals of the second kind. But then paths that represent irrationals converge just as much as these latter do, and so the critic's objection is really not clear.

[28] To elaborate: Let z_n denote the number represented in decimal by a single 1 followed by exactly n 6s. Then the limit of z_n as n goes to infinity is one sixth, because for all $\varepsilon > 0$ there exists an integer N such that if $n > N$, then the difference between z_n and one sixth is less than ε. But note carefully that the question "When does z_n become *equal to* one sixth?" makes no sense (how many 6s would you need?).

well? Irrationals and "rationals of the second kind" both require an infinite number of bits for their exact representation.

■ "But I don't think even the rationals can be counted using Sayan's method (I do know the method usually given for counting the rationals)." Now, this point I really don't understand. The critic has already said he "accept[s] that all of the rationals can be represented by Sayan's tree" (see the previous bullet item); so how can it be that Sayan's method of counting isn't sufficient to count them? To repeat, I don't really understand this claim; but to the extent I do understand it, it's surely wrong.

As for the critic's knowing "the method usually given for counting the rationals": Well, the method in question begins by conceptually laying out an infinite square array that looks like this:

```
1/1   1/2   1/3   1/4   1/5   ...
2/1   2/2   2/3   2/4   2/5   ...
3/1   3/2   3/3   3/4   3/5   ...
4/1   4/2   4/3   4/4   4/5   ...
5/1   5/2   5/3   5/4   5/5   ...
6/1   6/2   6/3   6/4   6/5   ...
. . . . .
```

As you can see, this array is defined very simply: What appears at the intersection of row i and column j is the fraction n/d, where $n = i$ and $d = j$. Clearly, this array does contain all possible rationals (in fact, it contains every one of those rationals an infinite number of times). Crucially, however, those rationals aren't represented (as they are in Sayan's tree) in terms of, e.g., conventional binary or conventional decimal notation—i.e., by means of strings of digits, all of which use some specific common radix. Instead, they're represented by vulgar fractions of the form n/d, where the numerator n and the denominator d are both decimal integers—and (in general) "improper" vulgar fractions at that, since n isn't constrained to be strictly less than d. So the question of whether some particular rational can or can't be represented using some specific radix simply doesn't arise. Thus, I don't see the relevance of the critic's point here—it strikes me as nothing but a red herring. Certainly it has nothing to do with Sayan's tree.

■ "I also accept that some irrational numbers (e.g., π) are defined by a convergent series, but Sayan claims that they can all be thus defined and I simply don't get it." But Sayan is right!—immediately and obviously right, I would have said, because the "convergent series" for a given irrational (assuming binary notation, of course) is, in effect, nothing but the bit string representation, and hence definition, of that irrational. That is, if the bit string in question is $t_1 t_2 t_3...$, then the irrational is defined to be equal to the sum

$$t_1/2 + t_2/4 + t_3/8 + ...$$

(once again I assume for simplicity that the irrational in question lies in the interval [0...1]).

Of course, it's true that some irrationals can also be represented by a rule, or formula; for example, there are many such formulas for π. But what's being claimed here isn't that *all* irrationals can be represented by some such formula[29]—it's merely that they can all be represented by some (infinite) string of bits, and hence by some (infinite) path in the tree.

Critic: All right, yes, I agree: Any given irrational can be represented as an infinite and convergent series, or in other words as a certain bit string. But we can't in general derive a precise limit from that convergence in the way we can for π.

Response: I don't know what the critic means here by that reference to "a precise limit for π." *No* irrational number has a value that's precisely representable in a finite number of digits (if that's what the critic means by his use of the term "precise limit" in this context). To any finite number of digits of precision the value of *any* irrational number is only ever an approximation, no matter what radix we happen to be using. (What's the precise value of π?) To say it one more time: The path in the tree for *any* number is just another way of writing out the binary representation of the number in question—e.g.,

```
1110010110000111...
```

(or whatever else it happens to be). The path has to be understood as going on forever, just as the normal way we write out the binary representation does. (Or as the normal way we write out the decimal representation does, come to that.)

Critic: I eventually realized that Sayan wasn't claiming that counting the nodes effectively counted the reals,[30] but rather that a certain property of all the finite trees carries through to the infinite one. The property in question is the fact—the assumption, rather—that the number of nodes in a finite tree is always less than the number of its paths; hence, I question the claim of countability that appears to depend on the opposite assumption.

Response: "The number of nodes in a finite tree is always less than the number of its paths"? No!—the number of nodes is always *equal* to the number of paths, by definition. In other words, I believe the critic's text here is muddled; to be specific, I believe by "paths" he means links and by "nodes" he means paths. If I'm right here, then the objection (paraphrasing somewhat as well) becomes:

> Sayan isn't claiming that counting the paths is equivalent to counting the reals, but rather that a certain property of every finite version of the tree carries through to the infinite case. The property in question is that the number of paths is less than the number of links. That property can't be true of the infinite tree; hence, I question the conclusion—viz., that the

---

[29] Whether they can or not is a separate question, one that I deliberately choose not to consider further here.

[30] Actually I think he was.

reals are countable—because that conclusion rests on the assumption that it *is* true.

Well, here's Sayan's actual text (but edited slightly to use my preferred terminology of links and paths, which Sayan calls segments and branches, respectively):

> I first take up the task of establishing that the real numbers in the interval [0...1) are denumerably infinite, by way of showing that the links can be paired off with natural numbers. This way of "counting" the links establishes that the set of links is denumerably infinite. I then prove that the set of paths of the tree, and thereby the reals in [0...1) they represent, cannot have a higher cardinality than the set of links of the tree. Hence the paths, and therefore the *reals* in [0...1) represented by the paths, can only be denumerably infinite.

> *Aside:* As you can see, Sayan considers the closed / open interval [0...1) rather than the closed / closed interval [0...1]. He also uses the term "denumerably infinite" instead of just plain "countable." However, these minor discrepancies are unimportant for present purposes. *End of aside.*

Contrary to what the critic asserts, Sayan carefully isn't claiming here that the number of paths (*nP*) is *less* than the number of links (*nL*); rather, he's claiming that the number of paths *nP* is *not greater* than the number of links *nL*, or equivalently that *nP* is less than *or equal* to *nL*. Of course, *nP* and *nL* both become infinite as the tree becomes infinite, and the notion of one infinity being less than another makes no sense. (Well, Cantor thought it did make sense—but Sayan doesn't, and neither do I.) Let me elaborate on this point. The critic goes on to assert (rather dogmatically, and without offering any supporting evidence) that the property "*nP* not greater than *nL*" can't remain true as *nP* and *nL* go to infinity. But there's a clear one to one into mapping (i.e., an injection) from the set of paths to the set of links!—and that mapping continues to hold as *nP* and *nL* go to infinity.[31] To spell out that mapping:

a. Each path is basically just a (nonempty) set of links.

b. No two distinct paths consist of the same set of links.

c. Hence, each distinct path can be paired in that mapping with the specific link (loosely, the "final" link in the path) that distinguishes the path in question from all the rest.

Given the existence of such a mapping, it follows immediately that *nP* can't be greater than *nL*. A little more formally:

---

[31] It's relevant to mention here that it was Cantor himself who first seriously relied on the idea of using mappings in such a context—and he did so precisely in order to get around the difficulties inherent in trying to compare the cardinalities of infinite sets (which is, of course, precisely the issue at hand). See Appendix B to this part of the essay.

- ■ We have two sequences of sets.

- ■ Call the sets in the first sequence P_1, P_2, ..., P_n, ... and the sets in the second sequence L_1, L_2, ..., L_n, ... *Note:* Informally, you can think of P_n as the set of paths to the nth level of the tree and L_n as the set of final links in the paths in P_n.

- ■ For all $n > 0$, let P_n and L_n have the same cardinality. *Note:* Given the interpretation of P_n and L_n suggested in the previous bullet item, that cardinality will in fact be 2^n.

- ■ Since P_n and L_n have the same cardinality, it follows that there must exist at least one mapping between them that's one to one and onto. *Note:* I talked above in terms of a one to one *into* mapping (not onto), but of course "one to one onto" is just a special case of "one to one into."

- ■ Let M_n denote this latter fact (i.e., the fact that such a mapping between P_n and L_n exists).

- ■ The critic claims that "nP not greater than nL" can't remain true as nP and nL go to infinity. That claim is equivalent to claiming there exists a concrete example of such sequences of sets P_1, P_2, ..., P_n, ... and L_1, L_2, ..., L_n, ... such that M_n ceases to hold as n goes to infinity.

Thus, if you believe the critic's claim here, it's only fair to ask you to provide such a concrete example. Personally, I don't believe any such example exists. In fact, it seems to me that to reject the idea that M_n continues to hold as n goes to infinity is to reject the entire corpus of (respectable!) mathematical argument dealing with infinity.

———— ◆◆◆◆◆ ————

To close out this appendix, here as promised is the critic's response to all of the foregoing.

Critic: The long story can be cut short: Sayan's paper shows no one to one correspondence between the real numbers and the integers, as is required to demonstrate countability. Instead, it attempts to prove just the *existence* of an enumeration, but I remain convinced that the attempt is phony. It relies on an assumption that the countability of the finite paths from the root node, representing numbers as denoted by a sequence of binary digits, is inherited by the infinite extensions of those paths.

I continue to maintain that Sayan's tree diverges, and because of that no conclusion based on a limit can be drawn about its infinite expansion. I also continue to maintain that only the rational numbers whose numerator is odd and whose denominator is a power of two are counted using Sayan's method. I remain supremely confident that the reals are uncountable.

(End of critic's response.)

APPENDIX D: LET HILBERT HAVE THE LAST WORD

The great mathematician David Hilbert (1862-1943) is widely believed to have agreed with, and indeed to have embraced, Cantor's ideas regarding infinity—the uncountability of the reals, the existence of an infinite number of distinct transfinite cardinals ($\aleph_0$, $\aleph_1$, $\aleph_2$, etc.), the behavior of the usual arithmetic operators ("+", "×", etc.) when dealing with infinite operands, and so on—in their entirety. Certainly the quote from Hilbert that I used as an epigraph to Part I of this essay lends credibility to this point of view. Here it is again:

> No one shall drive us out of the paradise that Cantor has created for us.

This quote is, of course, well known—even famous. As previously noted, it's from Hilbert's paper "On the Infinite" ("Über das Unendliche," *Mathematische Annalen 95*, 1926). But there are other remarks in that paper— quite a number of them, in fact—that seem to be much less well known, remarks that to my mind seriously undermine the general import of that quote, or even contradict it altogether. Let me illustrate.

First of all, here's the portion of Hilbert's text in which that quote actually appears (italics in the original): [32]

> The desires and attitudes which help us find [a way of avoiding paradox] are these ... Whenever there is any hope of salvage, we will carefully investigate fruitful definitions and deductive methods. We will nurse them, strengthen them, and make them useful. No one shall drive us out of the paradise that Cantor has created for us. Obviously, these goals can be attained only after we have fully elucidated *the nature of the infinite.*

So it seems to me that, given the context in which it appears (note in particular the final sentence of the text just quoted!), the idea that we can't and won't be driven out of "the paradise that Cantor has created for us" isn't to be taken as a categorical statement of fact. Rather, it seems to me much more in the nature of a desideratum on Hilbert's part; he's pointing out what we need to do if we're to have any hope of turning it into a true state of affairs. What it's clearly not—quite definitely not, it seems to me—is a definitive assertion, or claim, by Hilbert to the effect that "Cantor's paradise" is the way the world actually is.

---

[32] All otherwise unattributed quotes in this appendix are taken from a translation of this paper by Erna Putnam and Gerald J. Massey that was published in Paul Benacerraf and Hilary Putnam (eds.), *Philosophy of Mathematics: Selected Readings* (2nd edition, Cambridge University Press, 1984). But there's another Hilbert quote that I think is also worth repeating here (this one is from "Mathematical Problems," Hilbert's famous address to the International Congress of Mathematicians in Paris in 1900): "A mathematical theory is not to be considered complete until you have made it so clear that you can explain it to the first person you meet on the street." A good test to apply to Cantor's theory of the transfinite, I'd say. PS: I believe Hilbert actually said *man*, not *person*—but of course he lived in less enlightened times.

Still with reference to the text just quoted, I think there's more to be said. The fact is, I don't think it's even very clear just what Hilbert means by that phrase "the paradise that Cantor has created for us." Indeed, it seems to me entirely possible that he's referring not so much to the idea that a hierarchy of different infinities might exist (with all that such an idea entails), but rather to the idea that we need to reason about such matters very, very carefully—in other words, to the idea that we need to "carefully investigate," "nurse," "strengthen," and "make useful" those "fruitful definitions and deductive methods" that Cantor himself used.

For definiteness, however, let's agree for the remainder of this appendix that what Hilbert means when he refers to "Cantor's paradise" is indeed a world in which those different infinities do actually exist, with (to repeat) all that such an idea entails. Here then is Hilbert's very next paragraph in its entirety:

> We have already seen that the infinite is nowhere to be found in reality, no matter what experiences, observations, and knowledge are appealed to. Can thought about things be so much different from things? Can thinking processes be so unlike the actual processes of things? In short, can thought be so far removed from reality? Rather is it not clear that, when we think we have encountered the infinite in some real sense, we have merely been seduced into thinking so by the fact that we often encounter extremely large and extremely small dimensions in reality?

These words, again, hardly sound like someone claiming, or even wanting to claim, that "Cantor's paradise" is real. Au contraire, in fact: Hilbert seems to be stating here quite explicitly that "Cantor's paradise" is *not* the way the world actually is.

The following text, which appears in Hilbert's paper just prior to that remark about "the paradise that Cantor has created for us," also seems worthy of careful consideration (I've added some boldface):

> [Thanks] to the Herculean collaboration of Frege, Dedekind, and Cantor, the infinite was made king and enjoyed a reign of great triumph. In daring flight, the infinite had reached a dizzy pinnacle of success ... [But, in] the joy of discovering new and important results, mathematicians paid too little attention to the validity of their deductive methods. For, simply as a result of employing definitions and deductive methods which had become customary, contradictions began gradually to appear. These contradictions, the so called paradoxes of set theory, though at first scattered, became progressively more acute and more serious. In particular, a contradiction discovered by Zermelo and Russell[33] had a downright catastrophic effect when it became known throughout the world of mathematics. **Confronted by these paradoxes, Dedekind and Frege completely abandoned their point of view and retreated** ... Too many different remedies for the paradoxes were offered, and the methods proposed to clarify them were too variegated ... [The] present state of affairs where we run up against the paradoxes is intolerable. Just think, the definitions and deductive methods which everyone learns, teaches, and

---

[33] Here Hilbert is refrring to Russell's Paradox (see Part I, footnote 21).

uses in mathematics, the paragon of truth and certitude, lead to absurdities! If mathematical thinking is defective, where are we to find truth and certitude?

Here Hilbert is again very clearly recognizing the fact that Cantor's ideas lead to problems, and rather serious problems at that. So does he (Hilbert) solve those problems in his paper? To my way of thinking, no, he doesn't.

Finally, here's how Hilbert sums up his position on these matters in his paper's closing remarks (italics added for emphasis):

> In summary, let us return to our main theme and draw some conclusions from all our thinking about the infinite. *Our principal result is that the infinite is nowhere to be found in reality. It neither exists in nature nor provides a legitimate basis for rational thought*—a remarkable harmony between being[34] and thought. In contrast to the earlier efforts of Frege and Dedekind, we are convinced that certain intuitive concepts and insights are necessary conditions of scientific knowledge, and logic alone is not sufficient. *Operating with the infinite can be made certain only by the finitary.*
>
> The role that remains for the infinite to play is solely that of an idea—if one means by an idea, in Kant's terminology, a concept of reason which transcends all experience and which completes the concrete as a totality—that of an idea which we may unhesitatingly trust *within the framework erected by our theory.*

Well, I don't know about you, but to me this all looks very much like the words of a man who wants to have his cake and eat it too. Note in particular the apparent contradiction between (a) "the infinite [fails to] provide a legitimate basis for rational thought" and (b) "operating with the infinite can be made certain"! Though on balance—and to put the matter a little more positively, perhaps—it also looks to me as if Hilbert is at least admitting, perhaps a little regretfully, that the idea of "actual infinity" as opposed to "potential infinity" is nothing but a mind game after all.[35]

---

[34] Hilbert's actual word here was *Sein*, a word that in this context might better be translated as *existence*, or *actuality*, or even *reality*.

[35] Once again, see Appendix C to Part II of this essay if you need to refresh your memory regarding the logical difference between these two kinds of infinity.

www.ingramcontent.com/pod-product-compliance
Lightning Source LLC
Chambersburg PA
CBHW051758200326
41597CB00025B/4598

* 9 7 8 1 6 3 4 6 2 3 2 7 8 *